Lecture Notes in Networks and Systems 1557

The series “Lecture Notes in Networks and Systems” publishes the latest developments in Networks and Systems—quickly, informally and with high quality. Original research reported in proceedings and post-proceedings represents the core of LNNS.

Volumes published in LNNS embrace all aspects and subfields of, as well as new challenges in, Networks and Systems.

The series contains proceedings and edited volumes in systems and networks, spanning the areas of Cyber-Physical Systems, Autonomous Systems, Sensor Networks, Control Systems, Energy Systems, Automotive Systems, Biological Systems, Vehicular Networking and Connected Vehicles, Aerospace Systems, Automation, Manufacturing, Smart Grids, Nonlinear Systems, Power Systems, Robotics, Social Systems, Economic Systems and other. Of particular value to both the contributors and the readership are the short publication timeframe and the world-wide distribution and exposure which enable both a wide and rapid dissemination of research output.

The series covers the theory, applications, and perspectives on the state of the art and future developments relevant to systems and networks, decision making, control, complex processes and related areas, as embedded in the fields of interdisciplinary and applied sciences, engineering, computer science, physics, economics, social, and life sciences, as well as the paradigms and methodologies behind them.

Indexed by SCOPUS, EI Compendex, INSPEC, WTI Frankfurt eG, zbMATH, SCImago.

All books published in the series are submitted for consideration in Web of Science.

For proposals from Asia please contact Aninda Bose (aninda.bose@springer.com).

Paweł Trajdos · Robert Burduk
Editors

Progress in Pattern Classification and Machine Learning

Proceedings of the 14th International Conference on Computer Recognition Systems 2025

Springer

Editors
Paweł Trajdos
Wrocław University of Science
and Technology
Wrocław, Poland

Robert Burduk
Wrocław University of Science
and Technology
Wrocław, Poland

ISSN 2367-3370 ISSN 2367-3389 (electronic)
Lecture Notes in Networks and Systems
ISBN 978-3-032-01772-7 ISBN 978-3-032-01773-4 (eBook)
https://doi.org/10.1007/978-3-032-01773-4

This Springer imprint is published by the registered company Springer Nature Switzerland AG
The registered company address is: Gewerbestrasse 11, 6330 Cham, Switzerland

Preface

This volume of LNNS contains the proceedings of the 14th International Conference on Computer Recognition Systems (CORES 2025) held in Wrocław, Poland, (virtually) in May 2025.

The CORES conference brought together researchers, developers, practitioners, and educators in machine learning, computer recognition systems, and advanced computer systems.

The conference proceedings contain high-level papers selected through a strict review process.

Without high-quality submissions from the authors, the success of the conference would not be possible. Therefore, we would like to thank all authors and reviewers for their efforts in their submissions and evaluation.

Robert Burduk
Michał Woźniak
Marek Kurzyński
Conference Chairs

Organization

CORES

Honorary Chair

Juliusz Lech Kulikowski, Poland

Chairs

Michał Woźniak	Wroclaw University of Science and Technology, Poland
Robert Burduk	Wroclaw University of Science and Technology, Poland
Marek Kurzyński	Wroclaw University of Science and Technology, Poland

Contents

Suggested Future Trends in Computer-Aided Recognition Methods and Problems

Juliusz Lech Kulikowski(✉)

Institute of Biocybernetics and Biomedical Engineering, Polish Academy of Science, Warsaw, Poland
juliuszkulikowski@gmail.com

Abstract. The paper presents a general approach to computer-aided recognition problems as identification of relations satisfied by the objects under examination. Pattern recognition is considered as a binary relation of objects' similarity. Five other types of binary relations: matching, adjacency, distinction, ordering and attachability are described and illustrated as backgrounds of other types of recognition problems and procedures. An extension of the relations-based approach to recognition problems on higher-order relations, as well as on algebraic combinations of relations is described. The role of a proper similarity measure in solving a given relation recognition problem is described. Finally, a possibility of significant extension of the investigation area of computer-aided objects recognition is concluded.

Keywords: computer-aided recognition · pattern recognition · binary relations · similarity · matching · adjacency · distinction · ordering · attachability

1 Introduction

The year 1958, when the first monograph concerning a concept of "*Perceptron*" by F. Rosenblatt was published [9], is usually mentioned as a starting point of pattern recognition – a discipline aimed at creation and examination of the systems of artificial recognition of various objects. However, perceptron and its further versions represented only one of possible approaches to the pattern recognition problem. Almost simultaneously, a problem of radar echo-signals detection in noise arose and was investigated by advanced statistical methods [1,7]. Moreover, in the 60 s of the past century it occurred that artificial recognition of composed objects needs using more advanced methods, like based on formal grammars [3] or algebra of relations [4]. Currently, objects recognition is a discipline using various approaches: artificial neural networks, statistical, formal linguistic, algebraic, analytical etc. methods. However, a question concerning further development of this discipline arises. Of course, extension of the above

P. Trajdos and R. Burduk (Eds.): CORES 2025, LNNS 1557, pp. 1–8, 2025.
https://doi.org/10.1007/978-3-032-01773-4_1

methods on new classes of objects is possible and desired. Computer-aided recognition of various composite: biological, social, technological etc. objects and of their states will inspire to create new, more effective solution methods and algorithms. However, the aim of this presentation is to demonstrate that computer-aided recognition of any sort of *similarities* between objects is only one of the possible relationships that can be taken into consideration.

In general, an *n-th order relation* described on a given set Ξ of elements (objects) is defined as a subset of the Cartesian product Ξ^n, i.e. of the set of all possible n-long sequences of the objects belonging to Ξ. In particular, a *binary relation* is a set of ordered pairs of objects satisfying the conditions characterizing the given relation.

Similarity is a binary relation satisfying the conditions of: 1) *reflexivity* (any object α is similar to itself); 2) *symmetry* (if α is *similar* to β, then β is *similar* to α); 3) *transitivity* (if α is *similar* to β and β is *similar* to γ, then α is *similar* to γ). A set of all considered objects satisfying the relation of *similarity* is called a *similarity class.* Setting similarity classes is the first step to recognition problem formulation. This can be done in several ways: 1) by arbitrary objects selection and indication, 2) by using any physical, biomedical, chemical or any other tests, 3) by formal (geometrical, topological, statistical, algebraic, logical etc.) description of the objects. The recognition problem then takes one of two basic forms:

1. *detection*: there is given an object under observation and a similarity class of objects of particular interest – answer the question, whether the object belongs to the similarity class, or not;
2. *classification*: there is given an object under observation and a family of similarity classes – answer the question, to which similarity class the object belongs.

Sometimes, indicating more than one similarity class as the answer to classification problem is admitted (e.g. in a primary medical diagnosis). In any case a solution of the recognition problem consists in proving the fact that a similarity relation is (or is not) satisfied by a given object and the objects of the similarity class (or classes) under consideration. However, besides the *similarity* there are several other binary relations and a possibility of taking them as bases of some important types of decision problems will be considered. The paper does not present any results of a research project. It is aimed at presenting some general remarks about the possible future trends in computer-aided recognition methods and problems, as a starting point of a broad discussion.

2 Binary Relations Recognition Problems

Binary relations from the *reflexivity*, *symmetry* and *transitivity* points of view can be considered. This leads to the list of their possible types shown in Table 1.

Lack of any *reflexive* and *not-symmetrical* relations results from the fact that *reflexivity* is by definition also a *symmetric* property. The following examples should illustrate the above-given scheme.

Table 1. Types of binary relations

	Reflexive	Symmetrical	Transitive
Similarity	yes	yes	yes
Matching	yes	yes	no
Adjacency	no	yes	yes
Distinction	no	yes	no
Ordering	no	no	yes
Attachability	no	no	no

2.1 Example 1

Matching. A group of patients during a long time-interval has been hospitalized in a medical clinic. Any two patients can be called mutually *matched* if for at least one day they were staying in the same sick-room of the clinic. Two types of *matching classes* then can be taken into account. A *local matching class* can be defined as a set of all patients mutually matched due to the fact of their common staying during a certain time-interval in the same sick-room of the clinic. A *personal matching class* consists of all patients *matched* to a given patient during (but not obviously – simultaneously) all his hospitalization time. A matching recognition problem may then lie in finding the local and/or personal matching classes in order to investigate a process of infection propagation in the clinic.

Neighbourhood is another example of *matching.* Everybody is their proper neighbour (reflexivity); if A is a neighbour of B, then B is a neighbour of A (symmetry); if A is a neighbour of B and B is a neighbour of C then it does not mean that A is a neighbour of C (non-transitivity).

In a larger context examination of *matching classes* (may be an important step to investigation of hidden connections among the members of a (human, biological etc.) population.

2.2 Example 2

Adjacency. Assume that a set of archeologic remains is under consideration. It consists of irregular form elements of some broken clay pots, their original forms and number being unknown. The goal consists in reconstruction of the pots by correct selection and gluing together the corresponding, adjacent parts. For this, the elements have been numbered and their contours have been numerically described (say, by their Freeman codes [2]). A pair of elements is called satisfying the adjacency relation if some continuous parts of their contour codes are identical. Such elements are then numerically "glued" together by neglection of the common part of contour codes while the remaining parts of the contours are connected into an extended contour of the adjacent pair of elements. Continuation of this process leads to gradual "gluing" together the elements into larger parts of the objects under reconstruction. The process is continued numerically up to reaching satisfactory results of reconstruction of the pots. At the next step it may be physically repeated according to the numerical description of proposed mutual positions of adjacent elements. Similar example of reconstruction a document given in the form of its shreds was described in [6].

Detection of identical parts in the pairs of contours is a crucial problem influencing the effectiveness of the algorithms of combining together the adjacent elements. Another problem follows from the ambiguity of contour fragments selected for being glued. This may lead to a tremendous number of candidates to be "glued" to a given part of contour and a strategy of reducing this number may be necessary.

2.3 Example 3

Distinction. In a certain sense this relation can be considered as complementary to similarity. Assume, for example, that a great bison herd is under observation. From a biological point of view its members constitute a sort of similarity class. However, they also can be claimed *distinct* from each other if some of their additional individual features are taken into consideration. In particular, for distinction of the herd members they should be numbered, marked and put on a list. This may be a step to long-term monitoring of selected herd members' individual behaviour.

In a more general case, a computer-aided distinction recognition problem may consist in detection of specific objects hidden among a set of closely similar to it but different ones. Detection of defective products, typos, sick persons or animals etc. belong to the class of distinction recognition tasks, as well.

2.4 Example 4

Ordering. This means unambiguous arrangement of the elements of an enumerable or non-enumerable set, according to a certain criterion of their distinction. In monitoring the behaviour of an assumed perpetrator registered by several cameras distributed in a city area, their escape route can be reconstructed by correct *ordering* the time-sequence of records that show them. In this case, the *ordering* recognition should be closely combined with a *pattern recognition* procedure.

The ordering relation cannot be used directly to describe the case of ordering admitting repetition of objects or of some their sub-sequences. This case will be described in more detail in the next section of the paper.

2.5 Example 5

Attachability. This relation is called sometimes a *mereological* relation, existing between a composite object and its direct components. A relationship between an ancestor and their direct descendants maybe given as a proper example; lack of reflexivity, symmetry and transitivity properties in this case is evident.

Despite the fact that the *attachability* and *distinction* relations are both non-reflexive and non-symmetrical, the main difference between them consists in the lack of symmetry in the pairs of mutually *attached* objects, while it exists in the pairs of mutually *distinct* objects. In practice, a computer-aided *attachability* recognition may be useful, e.g., in control of automatic composite products assembling.

3 Extended Computer-Aided Recognition Problems

The above-presented examples indicate the possibility of extending the area of computer-aided recognition problems. Moreover, it should be remarked that binary relations are no more but a small part of relationships existing and observed in the real world. A general problem of recognition of the objects satisfying some relations, can therefore be formulated as follows:

1. There is given a non-empty set Ξ of elements (any distinguishable objects), called below an *observation space*;
2. There is taken into account a n-th order Cartesian product Ξ^n, i.e. a set of all possible n-long sequences of the elements of Ξ (called *syndromes*);
3. There is given a n-factors relation among the elements of Ξ, i.e. a non-empty subset $\boldsymbol{r} \subset \Xi^n$, $\boldsymbol{r}$ being characterized by a subset $\boldsymbol{C} \subset \boldsymbol{r}$ (a teaching set) of its exemplary syndromes or by analytical description of the **r**-syndromes characteristic properties;
4. There is given an object ξ, $\xi \in \Xi$; check its being a component of any syndrome of $\boldsymbol{r}$ and if so, present these syndromes.

Checking the fact mentioned above in the point 4), can be executed in several ways. In general, it depends on the nature of the objects under examination. Natural (i.e. physical, human, biological, social, etc.) objects should be, first, represented by their, formal characteristics, accessible by computer systems. Formal objects (numbers, geometrical objects, numerical vectors, ordered data sequences, graphs, etc.) can directly be subjected to checking procedures. This is possible if formal nature and structure of the observed object ξ and of the elements of Ξ are in mutual conformity. An example of conformity typical and important for pattern recognition takes place if ξ, u and the elements of Ξ can be considered as elements of a metric space. In this case a *distance measure* μ, such that for any ξ, u, $w \in \Xi$:

$$\begin{gathered} \mu(\xi, u) \geq 0, \quad \mu(\xi, u) = 0 \leftrightarrow \xi = u, \\ \mu(\xi, u) = \mu(u, \xi), \\ \mu(\xi, u) + \mu(u, w) \geq \mu(\xi, w), \end{gathered} \tag{1}$$

to any pair (ξ, u) can be assigned.

A distance measure can be established, in particular, in any multi-dimensional linear vector space. Then, the following exponential transformation of the distance measure into a *similarity measure* σ can be used [5]:

$$\sigma(\xi, u) = exp[-\mu(\xi, u)]. \tag{2}$$

This leads to the similarity measure satisfying the following conditions:

$$\begin{aligned} 0 \leq \sigma(\xi, u) \leq 1, \quad \sigma(\xi, u) = 1 \leftrightarrow \xi = u, \\ \sigma(\xi, u) = \sigma(u, \xi), \\ \sigma(\xi, u) \cdot \sigma(u, w) \leq \sigma(\xi, w). \end{aligned} \tag{3}$$

Using the above-described similarity measure σ to checking the fact of an object ξ being close or being a component of some syndromes of a given relation is limited by the requirement of formal homogeneity of the analysed objects' components.

3.1 Example 6

Assume, that Ξ consists of the strings of the following numerical data, describing the current health state of a patient: $\boldsymbol{x} = [u, v, t]$, where:
u denotes a systolic pressure in *hPa*,
v denotes a diastolic pressure in *hPa*,
t denotes a heart pulse in *beats/min.*

No, distance-based similarity measure of the strings of data, described above by (3), can be applied directly to the non-homogenous objects, like $\boldsymbol{x}$. The problem is that additions (as necessary in the calculations) of the values of pressure to the number of heart beats, even if mathematically admissible, are physically senseless.

The expression (3c) suggests a correct solution of the similarity measure problem in the case of non-homogenous data strings describing the objects under observation. For the data-strings of the form: $\xi = [\xi_1, \xi_2, \ldots, \xi_k]$, where $\xi_1, \xi_2, \ldots, \xi_k$ are some homogenous sub-strings of numerical data (and so is the string $\boldsymbol{u} = [u_1, u_2, \ldots, u_k]$), the similarity measure should have the form:

$$\sigma(\xi, \boldsymbol{u}) = \sigma_1(\xi_1, u_1) \cdot \sigma_2(\xi_2, u_2) \cdot \ldots \cdot \sigma_k(\xi_k, u_k). \tag{4}$$

The above presented similarity measure extends the possibility of relations recognition to the objects described by long non-homogenous data strings. On the other hand, the number of relations observed in real world is unlimited. In practice, we are able to distinguish and to characterize formally only some selected types of relations. However, the number of known and formally investigated relations can be increased by using the algebra of relations approach [4]. Any relation is a set of its syndromes (whatever their form should be). Therefore, the rules of classical set algebra [8] can be applied to them. This substantially extends the family of relations which can be taken into account as bases of relation recognition problems.

3.2 Example 7

Let Ξ denote an observation space and $\boldsymbol{r}_\alpha, \boldsymbol{r}_\beta$ be some syndromes described on Ξ. Then:

1. $\boldsymbol{r}_\gamma = \boldsymbol{r}_\alpha \cup \boldsymbol{r}_\beta$ denotes a relation consisting of all syndromes of $\boldsymbol{r}_\alpha$ **or** $\boldsymbol{r}_\beta$;
2. $\boldsymbol{r}_\gamma = \boldsymbol{r}_\alpha \cap \boldsymbol{r}_\beta$ denotes a relation consisting of all syndromes of $\boldsymbol{r}_\alpha$ **and** $\boldsymbol{r}_\beta$ **together**;
3. $\boldsymbol{r}_\gamma = \boldsymbol{r}_\alpha \setminus \boldsymbol{r}_\beta$ denotes a relation consisting of all syndromes of $\boldsymbol{r}_\alpha$ **but not of** $\boldsymbol{r}_\beta$;
4. $\boldsymbol{r}_\gamma = \boldsymbol{r}_\alpha \div \boldsymbol{r}_\beta$ denotes a relation consisting of all syndromes of $\boldsymbol{r}_\alpha$ **or** $\boldsymbol{r}_\beta$ **but not of both them together**;
5. $\boldsymbol{r}_\gamma = \neg \boldsymbol{r}_\alpha$ denotes a relation consisting of all syndromes **not satisfying** $\boldsymbol{r}_\alpha$.

The role of the above-described algebraic operations can be illustrated by the following example. Let Ξ denote a set of children described by their: a) *age*, b) *sex*, c) *height* and let the following relations on Ξ_2 be defined:

1. $\boldsymbol{r}_1$ – selection of all mixed (*boy, girl*) pairs (construction of a *similarity class*);
2. $\boldsymbol{r}_2$ – *ordering* the pairs according to common age of their members;
3. $\boldsymbol{r}_3$ – *ordering* the pairs according to the criterion of relative difference of the heights of their members.

This leads to a composite relation described by the expression:

$$\boldsymbol{r} = (\boldsymbol{r}_2 \cup \boldsymbol{r}_3) \cap \boldsymbol{r}_1. \tag{5}$$

It describes a relation whose syndromes have the form of two, ordered in different way sequences of heterogenous pairs of children. This can be used e.g. in final selection of the pairs admitted to a dance competition.

4 Conclusions

It has been shown that the computer-aided recognition problems can be interpreted as checking the fact of some relations being satisfied by the objects under examination. From this point of view, the pattern recognition problems are based on detection of the fact of the objects belonging to some, previously established classes of similarity. This concept can be extended on any other type of relations. In particular, the binary relations of *similarity*, *matching*, *adjacency*, *distinction*, *ordering* and *attachability* can be taken into account as bases of new relation recognition problems. This possibility can be extended by taking into account more sophisticated algebraic combinations of simple relations. As a result, the extended, relations-based recognition methods may be taken into consideration as powerful tools of various real world aspects description and investigation.

Acknowledgments. The author would like to cordially thank prof. Marek Kurzynski, Wroclaw University of Science and Technology, for being inspired by him to write this paper and for helping to prepare the manuscript for publication. He is also grateful to his daughter Renata Maksymowicz for her help to improve the manuscript linguistically.

References

1. Capon, J.: Nonparametric methods for the detection of signals in noise. Columbia University, Technical Report T-1/N, New York (1959)
2. Freeman, H.: A Review of Relevant Problems in the Process of Line-Drawing Data. Automatic Interpretation and Classification of Images. Academic Press, New York (1969)
3. Knoke, P.J., Wiley, R.G.: A linguistic approach to mechanical pattern recognition. Proc. IEEE Comput. **7**, 23–38 (1967)
4. Kulikowski, J.L.: Algebraic methods in pattern recognition. In: Courses and Lectures no 85, pp. 31 – 45. CISM, Udine (1971)
5. Kulikowski, J.L.: Pattern recognition. In: Biocybernetics and Biomedical Engineering. Biomedical Imaging, pp. 60–75. Academic Publisher EXIT, Warsaw (2003)
6. Kulikowski, J.L.: Objects' matching recognition problem and its applications. In: Computational Techniques and Algorithms for Image Processing, pp. 100–115. Lambert Academic Publishing (2009)
7. Meter, V., Middleton, D.: Modern statistical approaches to reception in communication theory. IRE Trans. **PGIT4**, 17–26 (1954)
8. Rasiowa, H., Sikorski, R.: The Mathematics of Metamathematics. PWN, Warsaw (1968)
9. Rosenblatt, F.: The perceptron, a probabilistic model for information storage and organization in the brain. Psychol. Rev. **65**, 86–97 (1958)

Formal Foundations of Vague Language Transduction for Syntactic Pattern Recognition

Mariusz Flasiński(✉) and Janusz Jurek

Information Technology Systems Department, Jagiellonian University, ul. prof. St. Ł ojasiewicza 4, 30-348 Cracow, Poland

Mariusz.Flasinski@uj.edu.pl

Abstract. A novel model of transduction of vague languages for syntactic pattern recognition is introduced in the paper. A methodological motivation of the research into the developing of the model, which has resulted from application-oriented requirements, is presented. The formal tool of generalized translation scheme for modeling modifications/enhancements of structural descriptions is proposed. The attributed programmed pushdown transducer for performing the generalized translation of vague languages is introduced.

Keywords: Syntactic pattern recognition · Vague languages · Syntax-directed transduction

1 Introduction

Structural patterns are usually treated as elements of formal string languages in syntactic pattern recognition, SPR, [3,16–18]. Then, generative grammars are used for modeling such languages, while formal automata are used as parsers of these languages for the recognition/interpretation of patterns. If patterns are noisy/ambiguous, then, apart from the filtering/smoothing of the underlying images during an image preprocessing stage, additional image enhancing techniques should be applied for structural patterns before they are parsed. Language transduction techniques, which are based on syntax-directed translation schemes and transducers (special types of formal automata with output) [22], constitute one of the most effective formal tool for this purpose [5,17,18].

DPLL(k) grammars [11] are assumed to be one of standard models in syntactic pattern recognition [16], because of two following reasons. The first reason, of the theoretical nature, is their big generative power. They generate the whole class of *CFLs* and the considerable subclass of *CSLs*, like: $L_1 = \{a^n b^n c^n : n \geq 0\}$, $L_2 = \{a^n b^m c^n d^m : n, m \geq 0\}$, $L_3 = \{a^{2^n} : n \geq 0\}$ [16]. The efficient parsing algorithm [11] of the $\mathcal{O}(n^2)$ time complexity has been defined for these grammars. Secondly, they were successfully applied for solving a variety of pattern

P. Trajdos and R. Burduk (Eds.): CORES 2025, LNNS 1557, pp. 9–19, 2025.
https://doi.org/10.1007/978-3-032-01773-4_2

recognition problems, e.g. for the recognizing of trend signals by real-time control systems [11], Auditory Brainstem Response audiometry [12], the analyzing of USG images for diagnosing congenital defects of fetal palates [15,20], the generating of electrical load forecast [14,15], the analysis of brain tumor images for postoperative radiation therapy in brain glioma patients [2].

During the research into the last mentioned application area, a specific requirement for the analysis of brain glioma images has appeared, namely the need to enhance image preprocessing methods for the delineating of target volume for postoperative radiation therapy. It means that the methodology based on vague *DPLL*(k) languages should be extended by developing the novel model of language transduction. The medical aspects of the problem, the application of the system implemented for the analysis of brain glioma images and the results concerning the improvement in the delineation quality have been presented in [2]. However, the model of the transduction of vague *DPLL*(k) languages has not been presented yet. The presentation of this original and unique model is the aim of the paper. Thus, the paper is of methodological/theoretical nature.

The application open problem mentioned is discussed in the succeeding section. The framework of vague *DPLL*(k) languages [15] used for the research is presented in Sect. 3. The generalized translation scheme for vague *DPLL*(k) languages for modeling structural descriptions which fulfill the application-oriented requirements is introduced in Sect. 4, whereas the attributed programmed pushdown transducer for performing the generalized translation of these languages, is presented in Sect. 5. The last section contains concluding remarks.

2 Methodological Motivation

A typical course of the development of a string syntactic pattern recognition model consists of the following stages [16]. Firstly, a standard model based on Chomsky grammars is defined for an analysis of unambiguous patterns. Secondly, extensions of the model for analyzing noisy/ambiguous patterns are constructed [16,17]. This model can be defined with the following enhancements of "pure" Chomsky grammars: stochastic grammars [31–33], fuzzy grammars [1,6,25] or extended grammars (and error-correcting parsing) [4,19,24].

For our syntactic pattern recognition model of *DPLL*(k) grammars and languages [11], we have recently proposed the novel approach based on vague *DPLL*(k) languages and the underlying multi-derivational parser that includes, as a generalized approach, all the above mentioned enhancements for noisy/ambiguous patterns (i.e. stochastic, fuzzy and error-correcting) [15]. This new approach has been already used in the research into the analyzing of USG images for diagnosing congenital defects of fetal palates [15,20], the generating of electrical load forecast [14,15] and the analysis of brain tumor images [2].

The method of the analysis of brain tumor images presented in [2] is used for the delineating of target volume for postoperative radiation therapy in brain glioma patients. Of course, such images have to be filtered/smoothed with the help of image processing techniques, before they are parsed/analyzed by the syntactic pattern recognition module. It turns out that in case of the task of the delineating of target volume for radiation therapy, apart from the use of image processing operations, some modifications of the form of a structural pattern should be performed. These modifications should be based not on standard image processing features (e.g. edges, blobs, ridges), but on structural descriptors (e.g. the ones defined by (sub)strings of elementary patterns (*structural primitives*)) which can be used for representing the knowledge of radiation oncologists. The requirement of the use of such knowledge is very important, since there are two contradictory key determinants in radiation therapy, namely: the effective damaging of cancerous tissues and, on the other hand, the sparing of normal tissues, which is especially crucial in case of brain tumors.

In case of such a problem, language transduction (translation) models are used in syntactic pattern recognition [5,17,18]. A language transduction model consists of two components: a syntax-directed translation scheme which is used for modeling string modification according to expert knowledge (it plays an analogous role to a rule-based system in AI), and a transducer (a specific kind of an automaton with output) [22] which is used for performing modification of one string to another according to this knowledge. In the following two sections, both formalisms for vague *DPLL*(k) languages are defined and presented for the first time (the medical aspects of the problem, the application of the system implemented for the analysis of brain glioma images and the results concerning the improvement in the delineation quality have been presented elsewhere [2].)

3 Preliminaries

As we have mentioned, the novel model of vague languages, which generalizes the known syntactic pattern recognition, SPR, approaches to analyze noisy/ambiguous patterns, has been introduced in [15]. Let us present the fundamental concepts of this model. Let Σ_T be an s-element set of terminal symbols.

Definition 3.1. *A vague primitive/terminal symbol over* Σ_T *is a vector*

$$\mathbf{v} = ((a_{k_1}, \mu_{k_1}), (a_{k_2}, \mu_{k_2}), \ldots, (a_{k_j}, \mu_{k_j})),\ 1 \leq j \leq s,$$

where $a_{k_1}, a_{k_2}, \ldots, a_{k_j} \in \Sigma_T$ *are different symbols,* $\mu_{k_1}, \mu_{k_2}, \ldots, \mu_{k_j}$ *are the measures ascribed to* $a_{k_1}, a_{k_2}, \ldots, a_{k_j}$*, correspondingly.*
Any pair (a_{k_c}, μ_{k_c})*,* $c = 1, \ldots, j$*, is called a unit of the vague primitive* $\mathbf{v}$*, where* a_{k_c} *is called a descriptor of* $\mathbf{v}$*,* μ_{k_c} *is called the attribute of* a_{k_c}*.*

In Definition 3.1 the type of a measure function μ is not determined. Thus, the measures used in SPR models for the analysis of noisy/ambiguous patterns (probabilistic measure in stochastic grammars, membership function in fuzzy grammars) are particular cases of μ. Moreover, the vectorial nature of μ allows us to generalize the error-correcting parsing model by the multi-derivational parsing in our model presented in [15].

A set of grammar terminal symbols (alphabet) for vague primitives will be further denoted with $\mathbf{\Sigma_T}$ (instead of Σ_T).

In our model languages/grammars/syntax analyzers can be enhanced by the use of attributes of primitives. We introduce such a semantic-syntactic extension according to a standard approach [21], which is used in syntactic pattern recognition [16,17]. Thus, for any terminal symbol a_{k_c} in Definition 3.1, apart from the attribute μ_{k_c}, which represents its vagueness (e.g. the probability measure), other attributes defining its semantic features (e.g. its length, its angle of inclination) can be defined. Therefore, a vague primitive/terminal symbol will be extended to be of the form $\mathbf{v} = ((a_{k_1}, \boldsymbol{\mu}_{\mathbf{k_1}}), (a_{k_2}, \boldsymbol{\mu}_{\mathbf{k_2}}), \ldots, (a_{k_j}, \boldsymbol{\mu}_{\mathbf{k_j}}))$, where $\boldsymbol{\mu}_{\mathbf{k_c}}$, $c = 1, \ldots, j$, is the (ordered) *set of attributes* of the descriptor a_{k_c} of the vague terminal symbol $\mathbf{v}$. The set of attributes of the vague terminal symbol $\mathbf{v}$ is denoted with $\mathcal{A}_{\mathbf{v}}$. $\mathbf{v}_\bullet\alpha$ denotes the attribute α of the vague terminal symbol $\mathbf{v}$, i.e. $\mathbf{v}_\bullet\alpha \in \mathcal{A}_{\mathbf{v}}$. D_α denotes the set of possible values for the attribute α.

Now, we can introduce the following definition.

Definition 3.2. *A vague word is a string* $\mathbf{w} = \mathbf{w}_1\mathbf{w}_2\ldots\mathbf{w}_n$, *where* $\mathbf{w}_i$, $i = 1, \ldots, n$, *is a vague primitive.*

In order to be consistent with the enhancement of our model introduced above, the notion of the set of attributes of a vague terminal symbol will be extended to for a vague word. If a vague word $\mathbf{w} = \mathbf{w}_1\mathbf{w}_2\ldots\mathbf{w}_n$, then $\mathcal{A}_{\mathbf{w}} = \mathcal{A}_{\mathbf{w}_1} \cup \mathcal{A}_{\mathbf{w}_2} \cup \ldots \cup \mathcal{A}_{\mathbf{w}_n}$ is called the set of attributes of $\mathbf{w}$.

Finally, we can define a vague grammar.

Definition 3.3. *Let* $G = (\Sigma_N, \Sigma_T, P, S)$ *be a generative grammar of the Chomsky model,* $\boldsymbol{\Sigma}_{\mathbf{T}}$ *be the set of vague terminal symbols defined for* Σ_T. *The grammar* $G' = (\Sigma_N, \boldsymbol{\Sigma}_{\mathbf{T}}, P, S)$ *is called the vague grammar corresponding to* G.

Now, we present the model of *DPLL(k)* grammars [11]. The *programmed grammars*, introduced by Rosenkrantz in 1969 [27], are a classical model of increasing a generative power of context-free grammars, *CFGs*. It is obtained by a control of a derivation process with fields containing indices of grammar productions. These fields can be defined for each *CFG* production. In the Rosenkrantz model the fields should be pre-defined for the grammar. Therefore, they are *static*. The concept of *dynamically programmed grammars* which allows us to define a *CFG* with fields, which can be filled *dynamically* during a derivation process, was introduced in 1995 [16]. Now, we can present the following definition.

Definition 3.4. *A dynamically programmed (context-free), DP, grammar is a quadruple*
$G = (\Sigma_N, \Sigma_T, P, S)$, *where*
Σ_N *is a set of nonterminal symbols,*
Σ_T *is a set of terminal symbols,*
P *is a set of* n *productions of the form:*
$p_i = (\pi_i, L_i, R_i, A_i, DCL_i)$, *in which*
$\pi_i : \bigcup_{k=1,\ldots,n} DCL_k \rightarrow \{TRUE,\ FALSE\}$ *is the predicate of applicability of* p_i,
$L_i \in \Sigma_N$ *and* $R_i \in \Sigma^*$ *are the left- and right-hand sides of* p_i, *respectively,*

A_i is the sequence of actions of a type: add, read, move which are performed over $\bigcup_{k=1,\ldots,n} DCL_k$,
DCL_i is the derivation control tape for p_i,
S is the start symbol (axiom), $S \in \Sigma_N$.

(L_i, R_i) is called the core of p_i. For each $p_i, p_j \in P$, $i \neq j$, the core of p_i is different from the core of p_j. The operation $add(k, m)$ writes the symbol m on the cell of DCL_k which is under the head. The operation $read(k)$ returns the value read by the head of DCL_k. The operation $move(k)$ moves the head of DCL_k right.

In order to construct an efficient parser [11], two concepts of: *DP* grammars and well-known parsable $LL(k)$ grammars [28] have been joined in order to define $DPLL(k)$ grammars in [11]. Let us present the following notions.

Let $G = (\Sigma_N, \Sigma_T, P, S)$ be a *DP* grammar, $\eta, \beta \in \Sigma^*$, and $\mid x \mid$ denote the length of a string $x \in \Sigma^*$. $FIRST_k(\eta)$ denotes a set of all the terminal prefixes of the strings of the length k (or of a length less than k, if a terminal string shorter than k is derived from η) that can be derived from η in the grammar G that can be formalized in the following way.

$$FIRST_k(\eta) = \{x \in \Sigma_T^* : (\eta \overset{*}{\Longrightarrow} x\beta \wedge \mid x \mid = k) \ \vee \ (\eta \overset{*}{\Longrightarrow} x \wedge \mid x \mid < k)\}.$$

Definition 3.5. *Let $G = (\Sigma_N, \Sigma_T, P, S)$ be a DP grammar, $\underset{core}{\overset{*}{\Longrightarrow}}$ denotes a sequence of derivation steps consisting in the applying of production cores only. G is called a DPLL(k) grammar iff the following two conditions are fulfilled.*

1. *For every two leftmost derivations in G*

$$S \overset{*}{\Longrightarrow} \alpha Z \theta \Longrightarrow \alpha\beta\theta \underset{core}{\overset{*}{\Longrightarrow}} \alpha x$$

$$S \overset{*}{\Longrightarrow} \alpha Z \theta \Longrightarrow \alpha\gamma\theta \underset{core}{\overset{*}{\Longrightarrow}} \alpha y,$$

where $\alpha, x, y \in \Sigma_T^*$, $\beta, \gamma, \theta \in \Sigma^*$, $Z \in \Sigma_N$, the following condition holds.

$$\text{If } FIRST_k(x) = FIRST_k(y), \text{ then } \beta = \gamma.$$

2. *For G there exists $m > 0$ such that for any leftmost derivation*

$$S \overset{*}{\Longrightarrow} \alpha Z \theta \overset{\sigma}{\Longrightarrow} \alpha\beta\theta \ ,$$

where σ is the string of the indices of productions applied, if $\mid \sigma \mid \geq m$ then the first symbol of $\beta\theta$ is terminal.

Finally, we can define vague $DPLL(k)$ languages.

Definition 3.6. *Let $\mathbf{\Sigma_T}$ be a set of vague terminal symbols, $G = (\Sigma_N, \mathbf{\Sigma_T}, P, S)$ be a vague DPLL(k) grammar. The vague DPLL(k) language generated by the grammar G is the set*

$$L(G) = \{\phi \in \mathbf{\Sigma_T}^* : S \overset{*}{\Longrightarrow} \phi\}.$$

The class of vague DPLL(k) languages is defined as follows

$$\mathcal{L}(X) = \{L : \exists \ G \text{ - vague } DPLL(k) \text{ grammar such that } L = L(G)\}.$$

4 Generalized Translation Scheme for Vague Languages

The model of vague *DPLL*(k) grammars [15] is the starting point for the transduction of vague *DPLL*(k) languages. A standard language transduction [5,22,29] consists of two components: (syntax-directed) *translation scheme* (*SDTS*) which is used for modeling string modification according to expert knowledge and a *transducer* which is used for performing modification of one string to another according to this knowledge. As we have mentioned in Sect. 2, in case of the application of vague *DPLL*(k) languages for the delineating of target volume for postoperative radiation therapy in brain glioma patients [2], the knowledge of radiation oncologists is used to reconcile the need of effective damaging of cancerous tissues with the requirement of sparing of normal tissues. According to this knowledge, a pattern representation of a brain glioma image obtained from the image preprocessing stage is to be modified to reconcile these two contradictory determinants. This modification is modeled by a language *translation scheme.*

In our model of vague *DPLL*(k) grammars, a novel type of a translation scheme had to be defined to take into account the following features. Firstly, the semantic nature of the model, which has been obtained in an analogous way as for semantic-syntactic translation schemes in [23,26]. Other two specific features, namely the vectorial nature of terminal symbols and the programmability of vague *DPLL*(k) grammars, had not been considered in the source literature till now. Thus, we have constructed a generalized translation scheme which includes all features mentioned. Let us introduce the following definition.

Definition 4.1. *A generalized translation schema, GTS, is a sextuple* $GTS = (N, \mathbf{\Sigma_T}, \mathbf{\Delta_T}, \mathcal{A}, R, S)$*, where*
N *is a finite set of nonterminal symbols,*
$\mathbf{\Sigma_T}$ *is a finite set of input vague terminal symbols,*
$\mathbf{\Delta_T}$ *is a finite set of output vague terminal symbols,*
$\mathcal{A} = \bigcup_{H \in (\mathbf{\Sigma_T} \cup \mathbf{\Delta_T})} \mathcal{A}_H$ *is a finite set of attributes,*
R *is a finite set of translation rules of the form:*
(π, R_{SYN}, R_{SEM}) *, where*
R_{SYN} *is the syntactic rule of the form:*
$Z \rightarrow X,\ Y$ *, in which*
$Z \in N$*,* $X \in (N \cup \mathbf{\Sigma_T})^*$*,* $Y \in (N \cup \mathbf{\Delta_T})^*$*, and the nonterminals in* Y *are a permutation of the nonterminals in* X*,*
$\pi : \mathcal{A}_X \rightarrow \{TRUE,\ FALSE\}$ *is the predicate of applicability of the translation rule,*
R_{SEM} *is the set of semantic rules defined as follows:*
let $X = X_1X_2 \ldots X_k$ *,* $X_i \in (N \cup \mathbf{\Sigma_T})$ *,* $i = 1, \ldots, k$ *and* $Y = Y_1Y_2 \ldots Y_p$ *,* $Y_r \in (N \cup \mathbf{\Delta_T})$ *,* $r = 1, \ldots, p$ *. let* $\mathcal{A}_X$ *be the set of all the attributes of the vague terminal symbols among* $X_1, X_2, \ldots, X_k$ *,*
if $Y_m \in \mathbf{\Delta_T}$*,* $m \in \{1, \ldots, p\}$ *and* $Y_m{\bullet}\alpha$ *is its attribute, then a semantic rule is the expression of the form*
$\alpha := f(\beta_1, \beta_2, \ldots, \beta_s)$*, where*

$\beta_1, \beta_2, \ldots, \beta_s \in \mathcal{A}_X$,
$f : D_{\beta_1} \times D_{\beta_2} \times \ldots \times D_{\beta_s} \ \rightarrow \ D_\alpha$ *is a function,*
S *is the start symbol (axiom),* $S \in N$.

Now, we can define a translation modeled by *GTS*. An instantaneous configuration of *GTS* is described with $(\kappa, \psi) \in (N \cup \mathbf{\Sigma_T})^* \times (N \cup \mathbf{\Delta_T})^*$, where the first element denotes a sentential form of the source language word during a derivation and the second element denotes a sentential form of the target language word during a derivation.

Definition 4.2. *Let T be a generalized translation scheme defined as in Definition 4.1. We denote* $(\gamma Z \nu, \eta Z \theta) \underset{T}{\longmapsto} (\gamma x \nu, \eta y \theta)$, $Z \in N$, $\gamma, x, \nu \in (N \cup \mathbf{\Sigma_T})^*$, $\eta, y, \theta \in (N \cup \mathbf{\Delta_T})^*$ *if and only if there exists* $(\pi, R_{SYN}, R_{SEM}) \in R$ *such that* R_{SYN} *is* $Z \rightarrow x, \ y \ , \ \ \pi$ *is TRUE, and the values of the attributes of* y *are computed according to* R_{SEM}. *We say that* $\eta y \theta$ *is directly generated from* $\gamma x \nu$ *by* T, *and we call such a direct generating a translation step in* T.

The reflexive and transitive closure of the relation $\underset{T}{\longmapsto}$ is denoted by $\overset{*}{\underset{T}{\longmapsto}}$.

Definition 4.3. *The translation modeled by the generalized translation scheme* T *is the set*
$\tau(T) = \{(\sigma, \omega) : (S, S) \overset{*}{\underset{T}{\longmapsto}} (\sigma, \omega), \ \sigma \in \mathbf{\Sigma_T}^*, \ \omega \in \mathbf{\Delta_T}^*\}$.

5 Attributed Programmed Pushdown Transducer for Vague Languages

A generalized translation scheme models a translation process. In order to perform translations from a vague *DPLL*(k) language to another one, we should define the underlying transducer. In the case of our model, the transductions should be conditional and after performing a syntactic part of a transduction the proper values of the attributes should be computed and assigned to both output terminals and stack symbols. Let us introduce the following definition.

Definition 5.1. *A (deterministic) attributed programmed pushdown transducer, APPDT is a nine-tuple*
$APPDT = (Q, \mathbf{\Sigma_T}, \Phi, \mathbf{\Delta_T}, \mathcal{A}, \delta, q_0, Z_0, F)$, *where*
Q *is a finite nonempty set of states,*
$\mathbf{\Sigma_T}$ *is a finite set of input vague terminal symbols,*
Φ *is a finite set of stack symbols,*
$\mathbf{\Delta_T}$ *is a finite set of output vague terminal symbols,*
$\mathcal{A} = \bigcup_{H \in (\mathbf{\Sigma_T} \cup \mathbf{\Delta_T} \cup \Phi)} A_H$ *is a finite set of attributes,*
$\delta : Q \times (\mathbf{\Sigma_T} \cup \{\lambda\}) \times \Phi \times \Pi \longrightarrow Q \times \Phi^* \times \mathbf{\Delta_T}^* \times AM$ *is the transduction mapping, in which*
AM *is the family of attributing mappings defined in the following way:*
if $\delta(q, X, Z, TRUE) = (q', \phi, Y, am^q_{X,Z})$,
then the set of attributing mappings for this transduction $am^q_{X,Z}$ *belongs to* AM

if and only if it is defined as follows:
for $\alpha \in \mathcal{A}_Y \cup \mathcal{A}_\phi$ *an attributing mapping is the expression of the form*
$\alpha := f(\beta_1, \beta_2, \ldots, \beta_s)$, *where*
$\beta_1, \beta_2, \ldots, \beta_s \in \mathcal{A}_X$,
$f : D_{\beta_1} \times D_{\beta_2} \times \ldots \times D_{\beta_s} \rightarrow D_\alpha$ *is a function,*
$q_0 \in Q$ *is the initial state,*
$Z_0 \in \Phi$ *is the initial stack symbol,*
$F \subseteq Q$ *is a set of final states.*

An instantaneous configuration of *APPDT* is described with $(q, \rho, \phi, \kappa) \in Q \times \mathbf{\Sigma_T}^* \times \Phi^* \times \mathbf{\Delta_T}^*$, where the first element denotes a state of the transducer, the second - a part of an input string which has not been analyzed till now, the third element - the content of the stack (the first symbol is the topmost symbol of the stack), and the fourth element - an output string emitted till now.

APPDT transduces an input string ζ in the following way. $(q_0, \zeta, Z_0, \lambda)$ is the initial configuration. The transducer reads the symbols of ζ one by one, and it makes the succeeding steps as follows.
$(q, ax, Z\phi, y) \vdash (q', x, \vartheta\phi, yr) \quad \Leftrightarrow \quad \delta(q, a, Z, TRUE) = (q', \vartheta, r, am^q_{a,Z})$,
where $q, q' \in Q$, $a \in \mathbf{\Sigma_T} \cup \{\lambda\}$, $x \in \mathbf{\Sigma_T}^*$, $Z \in \Phi$, $\vartheta, \phi \in \Phi^*$, $r, y \in \mathbf{\Delta_T}^*$.
$\vdash^*$ denotes the reflexive and transitive closure of the relation $\vdash$.

Let $APPDT = (Q, \mathbf{\Sigma_T}, \Phi, \mathbf{\Delta_T}, \mathcal{A}, \delta, q_0, Z_0, F)$ be given. For *APPDT* the underlying attributed programmed pushdown automaton is the attributed programmed pushdown automaton defined as follows.
$APPDA = (Q, \mathbf{\Sigma_T}, \Phi, \mathcal{A}, \delta', q_0, Z_0, F)$, where
$\delta'(q, a, Z, TRUE) = (q', \vartheta)$ if and only if $\delta(q, a, Z, TRUE) = (q', \vartheta, r, am^q_{a,Z})$ for some $r \in \mathbf{\Delta_T}^*$.

The transduction performed by *APPDT* can be defined as follows.
(1) The vague word $\mathbf{u}$ is the translation of the vague word $\mathbf{w} \in L(APPDA)$, where *APPDA* is the underlying automaton of *APPDT*, *by a final state* if and only if $(q_0, \mathbf{w}, Z_0, \lambda) \vdash^* (q_m, \lambda, \xi, \mathbf{u})$, $q_m \in F$.
(2) The vague word $\mathbf{u}$ is the translation of the vague word $\mathbf{w} \in L(APPDA)$, where *APPDA* is the underlying automaton of *APPDT*, *by the empty stack* if and only if $(q_0, \mathbf{w}, Z_0, \lambda) \vdash^* (q_m, \lambda, \lambda, \mathbf{u})$, $q_m \in Q$.

Definition 5.2. *Let APPDT be given. Let* $L \subseteq \mathbf{\Sigma_T}^*$ *be a language such that* $L = L(APPDA)$, *where APPDA is the underlying automaton of APPDT. The transduction performed by APPDT is a function*
$\mathcal{T}_{APPDT} : \mathbf{\Sigma_T}^* \rightarrow \mathbf{\Delta_T}^*$ *such that*
$\mathcal{T}_{APPDT} = \{(\mathbf{w}, \mathbf{u}) : \mathbf{w} \in \mathbf{\Sigma_T}^*, \mathbf{u} \in \mathbf{\Delta_T}^*$ *satisfy one of the conditions (1), (2)* $\}$.

6 Concluding Remarks

The novel model of the translation of formal languages consisting of: a generalized translation scheme and an attributed programmed pushdown transducer within the syntactic pattern recognition framework of vague *DPLL(k)* languages [15] has been presented in the paper. The development of this model has been motivated by the research into the application of vague $DPLL(k)$ languages for the analysis of brain tumor images for postoperative radiation therapy in brain glioma patients. Thus, the new methodological foundations which constitute the essential enhancement of this syntactic pattern recognition new paradigm [15] have been introduced in the paper, whereas the medical aspects of the application problem and the corresponding results have been presented elsewhere [2].

The model presented will be further developed in two directions. Firstly, it will be enhanced more by the use of an error-correcting parsing scheme [16,17]. Secondly, it will be extended for more complex patterns which are represented by graph structures. The efficient $ETPL(k)$ graph grammar-based syntactic pattern recognition model was proposed in [7]. This model is of the big generative power [10,16] and it has been used for many pattern recognition applications, e.g.: [8,9,13] (for a survey of its applications see [16]). Then, we will be able to construct of a hybrid two-level syntactic pattern recognition scheme.

Acknowledgments. The authors have no competing interests to declare that are relevant to the content of this article.

References

1. Asveld, P.: Fuzzy context-free languages (Part I and II). Theor. Comput. Sci. **347**, 167–213 (2005)
2. Bilski, M., Bułat, M., Peszyńska-Piorun, M., Baranowska, I., Fijuth, J., Kuncman, Ł., Jurek, J.: . Computer aided target volume delineation for postoperative radiation therapy in brain glioma patients with the use of the hybrid Artificial Intelligence model. In: Karamarković (ed.) International Conference on Radiation Applications in Physics, Chemistry, Biology and Medical Sciences, p. 174. Sievert Publ., University of Granada, Granada, Spain (2024)
3. Bunke, H., Sanfeliu, A. (eds.): Syntactic and Structural Pattern Recognition - Theory and Applications. World Scientific, Singapore (1990)
4. Chan, K.F., Yeung, D.Y.: Error detection, error correction and performance evaluation in on-line mathematical expression recognition. Pattern Recogn. **34**, 1671–1684 (2001)
5. Fan, T.I., Fu, K.S.: A syntactic approach to time-varying image analysis. Comput. Graph. Image Process. **11**, 138–149 (1979)
6. Ferreira, M., Santos, C., Monteiro, J.: Cork parquet quality control vision system based on texture segmentation and fuzzy grammar. IEEE Trans. Ind. Electron. **56**, 756–765 (2009)
7. Flasiński, M.: Parsing of edNLC-graph grammars for scene analysis. Pattern Recogn. **21**, 623–629 (1988)
8. Flasiński, M.: Distorted pattern analysis with the help of nodel label controlled graph languages. Pattern Recogn. **23**, 765–774 (1990)
9. Flasiński, M.: On the parsing of deterministic graph languages for syntactic pattern recognition. Pattern Recogn. **26**, 1–16 (1993)

10. Flasiński, M.: Power properties of NLC graph grammars with a polynomial membership problem. Theor. Comput. Sci. **201**, 189–231 (1998)
11. Flasiński, M., Jurek, J.: Dynamically programmed automata for quasi contexts sensitive languages as a tool for inference support in pattern recognition-based real-time control expert systems. Pattern Recogn. **32**, 671–690 (1999)
12. Flasiński, M., Reroń, E., Jurek, J., Wójtowicz, P., Atłasiewicz, K.: On the construction of the syntactic pattern recognition-based expert system for Auditory Brainstem Response analysis. In: Kurzyński, M., Puchała, E., Woźniak, M., Żołnierek, A. (eds.) CORES 2005. Advances in Soft Computing, vol. 30, pp. 503–510, Springer, Heidelberg (2005). https://doi.org/10.1007/3-540-32390-2_59
13. Flasiński, M., Myśliński, S.: On the use of graph parsing for recognition of isolated hand postures of Polish Sign Language. Pattern Recogn. **43**, 2249–2264 (2010)
14. Flasiński, M., Jurek, J., Peszek, T.: On the learning of vague languages for syntactic pattern recognition. Pattern Anal. Appl. **26**, 605–615 (2023)
15. Flasiński, M., Jurek, J., Peszek, T.: Multi-derivational parsing of vague languages - the new paradigm of syntactic pattern recognition. IEEE Trans. Pattern Anal. Mach. Intell. **46**, 5677–5691 (2024)
16. Flasiński, M.: Syntactic Pattern Recognition. World Scientific, New Jersey-London-Singapore (2019)
17. Fu, K.S.: Syntactic Pattern Recognition and Applications. Prentice Hall, Englewood Cliffs (1982)
18. Gonzales, R.C., Thomason, M.G.: Syntactic Pattern Recognition: An Introduction. Addison-Wesley, Boston (1978)
19. Huang, K.Y.: Syntactic Pattern Recognition for Seismic Oil Exploration. World Scientific, New Jersey-London-Singapore (2002)
20. Jurek, J., Wójtowicz, W., Wójtowicz, A.: Syntactic pattern recognition-based diagnostics of fetal palates. Pattern Recogn. Lett. **133**, 144–150 (2020)
21. Knuth, D.: Semantics of context-free languages. Math. Syst. Theory **2**, 127–145 (1968)
22. Lewis, P.M., II., Stearns, R.E.: Syntax-directed transduction. J. ACM **15**, 465–488 (1968)
23. Lewis, P.M., Rosenkrantz, D.J., Stearns, R.E.: Attributed translations. J. Comput. Syst. Sci. **9**, 279–307 (1974)
24. Lopez, D., Sempere, J.M., Garcia, P.: Error correcting analysis for tree languages. Int. J. Pattern Recogn. Artif. Intell. **14**, 357–368 (2000)
25. Mordeson, J.N., Malik, D.S.: Fuzzy Automata and Languages. Theory and Applications. Chapman and Hall - CRC, Boca Raton (2002)
26. Pyster, A., Buttelmann, H.W.: Semantic-syntax-directed translation. Inf. Control **36**, 320–361 (1978)
27. Rosenkrantz, D.J.: Programmed grammars and classes of formal languages. J. ACM **16**, 107–131 (1969)
28. Rosenkrantz, D.J., Stearns, R.E.: Properties of deterministic top-down grammars. Inf. Control **17**, 226–256 (1970)
29. Thomason, M.G.: Stochastic syntax-directed translation schemata for correction of errors in context-free languages. IEEE Trans. Comput. **24**, 553–572 (1975)
30. Thomason, M.G.: Stochastic syntax-directed translations in syntactic pattern recognition. Pattern Recogn. **14**, 187–190 (1981)
31. Tsai, W.H., Fu, K.S.: A pattern deformational model and Bayes error-correcting recognition system. IEEE Trans. Syst. Man Cybern. **9**, 745–756 (1979)

32. Vidal, E., Thollard, F., Higuera, C., Casacuberta, F., Carrasco, R.: Probabilistic finite-state machines (Part I and II). IEEE Trans. Pattern Anal. Mach. Intell. **27**, 1013–1039 (2005)
33. Zhu, S.C., Mumford, D.: A stochastic grammar of images. Found. Trends Comput. Graph. Vision **2**, 259–362 (2007)

A Target-Oriented Image Fusion Method for Quality Inspection of Heat-Sealed Packages

Piotr Garbacz[1,2](✉)

[1] Łukasiewicz Research Network – Institute for Sustainable Technologies, Radom, Poland
[2] Doctoral School of the Rzeszow University of Technology, Rzeszów, Poland
piotr.garbacz@itee.lukasiewicz.gov.pl

Abstract. Quality inspection is recognized as a critical aspect of manufacturing, profoundly affecting product integrity and customer satisfaction. The paper presents the application of the multimodal imaging technique, specifically infrared (IR) and visible light (VIS) imaging, that is investigated to enhance the inspection of heat-sealed packages. The distinct strengths of each imaging modality are leveraged to address the limitations associated with single-sensor approaches. An effective image fusion algorithm is proposed to integrate complementary information, facilitating the detection of subtle defects in heat-sealed packages, such as inconsistencies in seals and contamination. Experimental results show that the proposed method enhances defect classification and overall inspection accuracy by synthesizing redundant and complementary data sources.

Keywords: Image fusion · Multimodal imaging · Quality inspection

1 Introduction

In the relentless pursuit of manufacturing excellence, quality control stands as a critical pillar. Significant advancements in inspection technologies, specifically the implementation of infrared [1], x-ray [2], terahertz [3] and hyperspectral imaging [4], are providing manufacturers with enhanced analytical capabilities. These technologies, which have experienced substantial adoption in recent years, offer unparalleled insights into product integrity, enabling the identification of defects undetectable through conventional visual inspection. The widespread adoption of microbolometer sensors has facilitated the integration of thermal imaging into diverse industrial quality control and diagnostic systems. Thermal imaging offers a non-destructive analysis of temperature distributions, enabling the detection of anomalies indicative of defects in various materials and processes. Furthermore, developing compact terahertz radiation sources and sensors has unlocked new applications for terahertz technology within industrial environments. Hyperspectral imaging (HSI) systems enable the registration of several

P. Trajdos and R. Burduk (Eds.): CORES 2025, LNNS 1557, pp. 20–29, 2025.
https://doi.org/10.1007/978-3-032-01773-4_3

hundred images in various bands of the electromagnetic spectrum. Spectral signatures are recorded for each pixel in the image of the observed surface, showing the relationship between the light reflectance and the wavelength. HSI has so far been used mainly in laboratories and off-line analysis systems. However, due to the development in hyperspectral sensing technologies, it is increasingly used in on-line quality inspection on production lines [5]. The limitations inherent in single-measurement inspection techniques have spurred significant research and development in multi-sensor systems. Next-generation systems are being developed to enhance automation, accuracy, and reliability. These limitations of single-modality inspection systems, such as sensitivity to specific defect types or environmental conditions, drive the need for more robust solutions. Hybrid, or multimodal imaging [6], which combines multiple techniques and employs image fusion, represents a promising approach. By integrating data from different sensor modalities, hybrid imaging systems can provide a more comprehensive view of the inspected object. For example, fusing thermal and visual data can enhance defect detection by combining information about temperature variations with surface features [7]. Acquiring and processing data from multiple sources in real-time enables proactive identification of quality issues, optimized process control, and reduced waste. The ongoing research in this area [8], including optimizing sensor selection and integration, developing robust image fusion algorithms, and ensuring the real-time performance of these systems in demanding industrial environments. Further research is needed to develop standardized frameworks for evaluating hybrid inspection systems performance and address the computational complexity associated with multimodal data processing [9].

2 Hybrid Inspection Method for Heat-Sealed Packaging

Ensuring the quality and integrity of heat-sealed packaging is critical in various industries, particularly food and pharmaceuticals. Traditional quality control methods often struggle to detect the full spectrum of potential defects, necessitating the exploration of more comprehensive approaches. Defects that arise after the packaging process can stem from several factors, including incorrect time and temperature settings, contamination of the sealing area, or improper positioning of components. The innovative aspect of the research is the potential to utilize information about classified defects—obtained through advanced imaging—for use in a predictive quality system that leverages machine learning to anticipate and prevent quality issues [18]. The proposed method leverages the strengths of different sensor modalities to provide a more holistic assessment of package integrity. Classical vision systems excel at detecting mechanical defects such as leaks, cracks, and package deformation. Thermal imaging provides insights into the temperature distribution within the sealing zone, enabling the identification of defects stemming from suboptimal sealing parameters. The utilization of a multi-modal approach demonstrates significant potential for enhancing the precision and resilience of heat-sealed packaging inspection, while also fostering synergistic effects. For the inspection of heat-sealed packages, a hybrid method offers the advantage of:

- increasing the effectiveness and reliability of surface defect detection;
- improving the quality of the seal and the durability of the packaging;
- reducing food waste by early detection of defects and preventing the need to dispose of entire batches of products;
- optimizing the sealing process by identifying the causes of defects.

Hybrid inspection systems for heat-sealed packaging are a promising solution that can contribute to improving the quality and safety of food products and reducing losses in the food industry. However, effective integration of data from diverse sources requires two critical steps: image registration and fusion. To develop model algorithms for image fusion and defect detection, a laboratory set-up was created that includes both a thermal imaging camera and a visible band camera (see Fig. 1). The stand enables the transport of packages at a set speed of up to 1 m/s.

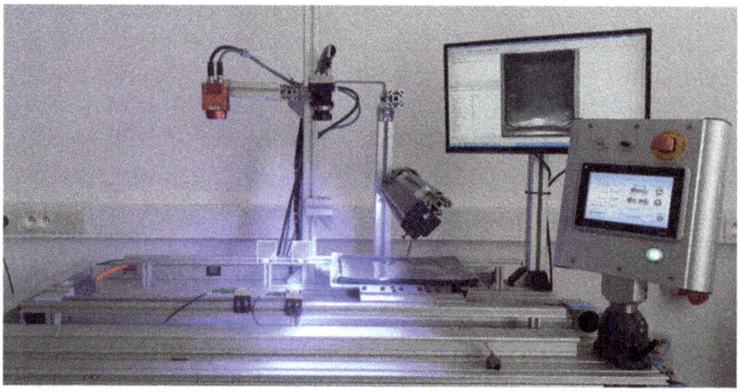

Fig. 1. Laboratory test stand for IR-VIS quality inspection.

To gather a representative set of images, defects that can occur during the packaging process were simulated by placing poultry meat samples in the heat sealing area, unevenly arranging the packaging film, and adjusting the temperature and duration of the sealing process. Every test carried out was documented through a comprehensive metadata file. The data structure included critical information such as acquisition time, duration of the sealing process, and specified sealer temperature. The rapid cooling process of the package presented a significant challenge in maintaining consistent acquisition times for the infrared (IR) camera. Following the identification and removal of outliers resulting from operator errors during the experiment, the data set has been refined to consist of 125 image pairs. The average time to acquire data from the thermal imaging camera after the sealing process was 8.67 s, with a standard deviation of 0.02 s. According to the manufacturer's guidelines, the standard parameters for the packaging materials and foil used in the tests require a sealing temperature of 190°C for a time of 2 s. The average sealing time for all samples was 2.12 s, with a standard deviation of 0.3 s. Furthermore, the samples were classified according to their heating temperatures: nominal temperature with 80 samples; low temperatures of 170°C (10 samples) and 180°C (5 samples); and high temperatures of 200°C (5 samples), 210°C (5 samples) and 220°C (20 samples).

3 Target-Oriented IR-VIS Fusion Method

In industrial inspection systems, image fusion should serve as a reliable data source for analysis algorithms while providing a clear and intuitive visual representation to help human operators identify the source of defects. In addition, in many applications, the processing time of implemented algorithms is limited, which necessitates the pursuit of target-oriented robust solutions. To effectively integrate the high spatial frequency of surface texture while maintaining a prominent infrared target [10], the fusion methodology must ensure a nondestructive combination of these information modalities. This is particularly challenging when dealing with modalities that have significantly different characteristics. The proposed method employs fused pseudo-RGB images generated from the simultaneous use of thermal and visible cameras. The prevalent paradigm in infrared and visible image fusion is characterized by a reliance on global image processing, wherein fusion algorithms operate on aggregated features derived from the entirety of the image scene [11]. This global approach inherently reduces the preservation of fine-grained target-specific information. Fusion methods can generally be categorized into several groups, including [12] multiscale transform, sparse representation, neural network and saliency-based methods, and hybrid models. Methodologies that explicitly address target-level fusion, incorporating techniques such as localized feature extraction and region-based fusion strategies, constitute a minority within the current state of the art.

Traditional inspection systems are typically designed for specific quality control tasks and usually focus on a single object type within strictly regulated environments. [13]. This allows for accurate calibration, consistent lighting, and predictable positioning of objects, which in turn achieves reliable and repeatable measurements. Incorporating a priori knowledge concerning object and process variability facilitates the implementation of computationally efficient and robust machine vision algorithms. This approach mitigates algorithmic complexity without compromising inspection reliability and improves system adaptability in real-world manufacturing scenarios. In accordance with these principles, an image fusion algorithm, tailored for the analysis of the sealing region, was developed. (see Fig. 2). The initial phase of the image registration involves pre-processing both infrared (IR) and visible (VIS) images. In the IR domain, the principle that the highest temperature concentration signifies the heating area's center is leveraged. The significant distortion inherent in thermal camera lenses necessitates image rectification [14]. This is achieved using a pre-generated rectification map, derived from a calibration target detectable in both infrared and visible spectra. Subsequently, the heating path's central axis is determined through the application of a ridge detection algorithm [15]. In the VIS domain, traditional edge detection is utilized, capitalizing on the high contrast between the bright background and the dark tray. It is worth pointing out that, due to positioning variations during the sealing process, the sealing area's location can only be approximated based on outline of the detected object. To accurately determine the sealing area's center within the visible image, further analysis is required (see Fig. 3 a). Before the analysis, the image undergoes negation, followed by morphological operations to eliminate extraneous artifacts and isolate the sealing area. Subsequently, a stripe detection algorithm [15] is applied within predefined regions of interest. Although the pre-processing filtering stage effectively reduces noise, it doesn't eliminate all unwanted edge peaks in the edge derivative response profile (see Fig. 3 b). Therefore, to ensure accurate stripe detection, parameters such as the minimum magnitude threshold and the allowable stripe width, are precisely configured. From the detected stripe centers, four lines are generated via interpolation, representing the tray's top, right, bottom, and left sealing areas. The intersections of these lines yield four corner points. The same procedure is then applied to the IR image, producing corresponding corner points that are subsequently used for image registration [16]. In the context of a close, planar scene and effective distortion removal during pre-processing, a perspective projection model was employed for global image mapping registration. This model depends on four pairs of corresponding points and guarantees the preservation of straight lines in the registered images [17]. Following registration, the aligned images are utilized to generate a pseudo-RGB image. Subsequently, a novel fusion architecture is employed to synthesize information from a thermal image and two monochromatic visible images: raw and contrast-enhanced in the sealing area (see Fig. 4). This fusion process yields an image that effectively encapsulates thermal radiation data from the infrared spectrum and fine textural details from the visible domain. The Fig. 4 illustrates three distinct regions of interest (ROIs) representing varying sealing quality. The top ROI depicts an area characterized by optimal sealing, evidenced by consistent temperature distribution and the absence of surface defects. Conversely, the middle-left ROI exhibits characteristics indicative of insufficient sealing temperature and, as a result, the presence of a weak bond. The bottom ROI, while demonstrating appropriate sealing temperature, reveals surface irregularities in the foil, manifesting as creases. To enhance defect visualization, a pseudo-RGB fusion technique was implemented. Thermal data was assigned to the red channel, leveraging the human perception of red as indicative of temperature. Visible light data was mapped to the green channel, and a contrast-enhanced image to the blue channel. This approach was particularly effective in highlighting low-temperature defects, where the reduced red channel intensity facilitated clear visual differentiation of defect types.

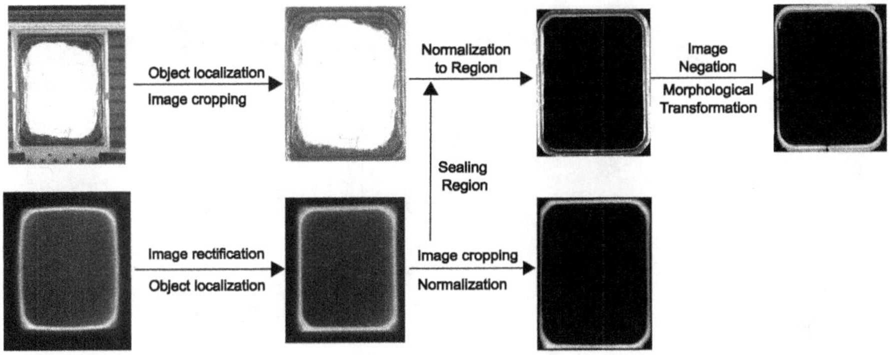

Fig. 2. Pre-processing stage for Infrared-Visible image registration.

a) 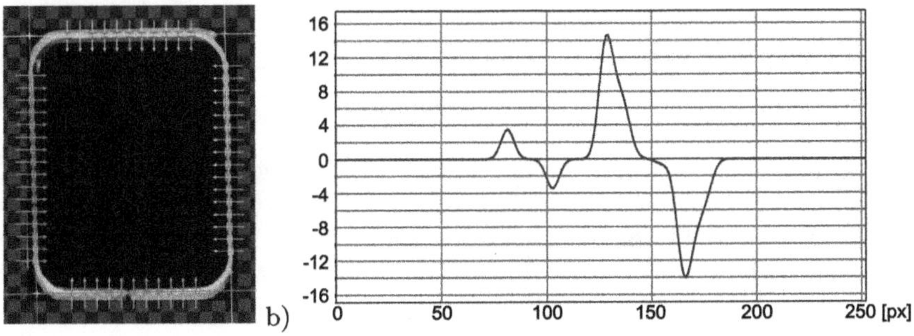

Fig. 3. Stripe detection algorithm a) sample VIS image b) example profile of the edge (derivative) operator response.

4 Experimental Results

The main objective of the experimental study was to evaluate the proposed concept of the IR-VIS inspection method for defect detection and classification. The tests were conducted under conditions that closely mimic industrial environments. Upon reviewing the samples, the defects identified in the sealing area included creases, wrinkles, contamination, and poor film bonding. Following excluding unsuitable samples, the final dataset consisted of 125 image pairs. The resolution of the visible (VIS) image was 3500×2200 pixels, while the infrared (IR) image had a resolution of 640×480 pixels. The tests were conducted on an industrial computer equipped with a 7th generation CPU running at 3 GHz.

4.1 Image Registration and Fusion

The transformations observed between the recorded sample images are predominantly translational and rotational, exhibiting no significant nonlinear deformations. As a result, the Euclidean distance metric has been used, as it is well known

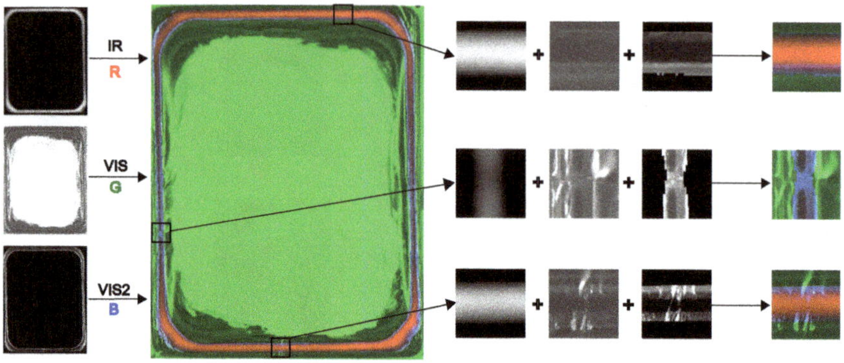

Fig. 4. Fusion architecture of pseudo-RGB image.

for its simplicity and effectiveness in measuring spatial differences between corresponding points, particularly in IR-VIS image fusion [19]. For the evaluation of the proposed algorithms, four corners of the sealing area were selected as they represent the alignment quality of the entire inspection region (Table 1). Considering that the calculated image scale factor is approximately 0.15 mm/px, the reported error is within an acceptable range, as the misalignment typically does not exceed 1 mm. Furthermore, due to time constraints in industrial inspection systems, it was essential to verify the execution time of the algorithm. For the selected sample group, the average total time for image registration and fusion was 50 ms, with a standard deviation of 5 milliseconds.

Table 1. The alignment accuracy using the Euclidean distance metric.

Distance	Mean [px]	Std dev. [px]	Maximum [px]
Top-right corner	0.92	0.77	4.43
Top-left corner	1.19	1.3	7.2
Bottom-right corner	0.99	0.74	4.47
Bottom-left corner	1.22	0.95	5.45

4.2 Defects Detection and Classification

To evaluate the effectiveness of the fused images in detecting various types of defects, preliminary testing utilizing deep learning was performed. Assessment was carried out using the Aurora Deep Learning development platform [20]. In the study, a feature detection tool (segmentation) was used, which enables the accurately labeling of each pixel corresponding to defects (see Fig. 5). In the training phase, the model is optimized to precisely segment classes of pixel-wise features within an image. The result is an array of probability heatmaps for every class. The data set used in the model training process was partitioned using the holdout method with a random split into three subsets (60-20-20): a training set containing 75 images, a validation set consisting of 25 images, and a test set comprising 25 images. The validation set was used for model optimization, while the test set was reserved for final performance evaluation. In addition, commonly employed augmentation techniques were implemented, including adjustments to brightness and contrast, blurring, random angle rotation, and both horizontal and vertical flips. Model performance was evaluated using the F1 score metric, and precision and recall were additionally calculated (Table 2). During network training, a divergence in loss trends was observed. The training loss continued to decrease gradually, while the validation loss remained stable at a constant level. This pattern, along with a significant gap between the performance of the training data and the unseen data, suggests potential overfitting, likely due to a limited number of sample images. Despite being at an early stage of research, the model's performance indicates that the generated fused images provide valuable input for not only defect detection but also detailed classification. Future research will explore various deep learning architectures using a larger sample set collected from a factory production line.

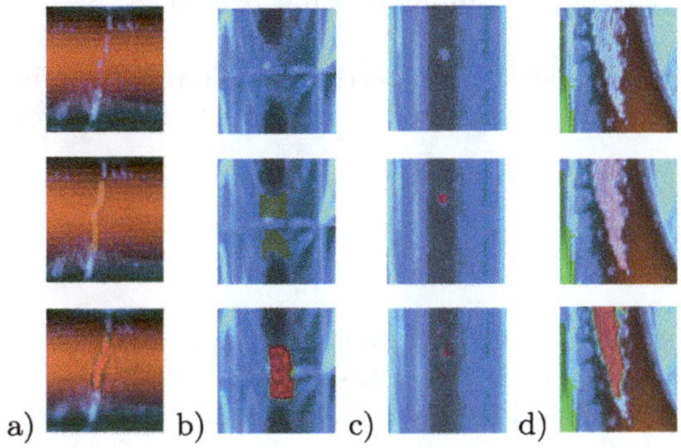

Fig. 5. Test set example images: pseudo-RGB (top), labeled (middle) and with heatmap (bottom) a) crease b) poor bond c) contamination d) wrinkles.

Table 2. Summary of evaluated segmentation model.

	Training set	Test set
F1 Score [%]	86.42	67.90
Recall [%]	89.36	65.68
Precision [%]	83.66	70.28

5 Summary

The assurance of product integrity and the preservation of consumer trust are contingent upon the implementation of rigorous quality control procedures within manufacturing processes. The image fusion method proposed in the paper demonstrates the advantages of combining infrared (IR) and visible light (VIS) imaging to improve the inspection of heat-sealed packages. By leveraging the unique characteristics of electromagnetic waves within specific optical radiation bands, this proposed technique creates redundant and complementary data through effective image fusion, enabling the preservation of key features. Robust image registration is essential, particularly for multi-modal fusion in industrial applications. Empirical findings confirm that the proposed technique significantly enhances flaw detection and overall inspection accuracy compared to conventional single-image methods. Future work will leverage the developed system as a data source for training predictive quality machine learning models. This predictive quality feedback loop will enable manufacturers to continuously improve quality outcomes. Predicted or detected defects will be subject to timely verification through root cause analysis. The integration of multimodal machine vision systems and machine learning facilitates proactive process adjustments, leading to substantial reductions in waste and downtime, and a fundamental transition towards predictive manufacturing practices.

Acknowledgments. This study was partially funded by the Ministry of Science and Higher Education as a part of the Implementation Doctorate Programme.

References

1. Venger, Y.F., et al.: Infrared thermography as an effective tool for research and industrial application. Sci. Innov. **17**(5), 20–33 (2021)
2. Michalska, M.: Overview of the use of X-Ray equipment in electronics quality tests. Informatyka, Automatyka, Pomiary w Gospodarce i Ochronie Środowiska **11**(2), 26–29 (2021)
3. Ellrich, F., Bauer, M., Schreiner, N., et al.: Terahertz quality inspection for automotive and aviation industries. J. Infrared Millimeter Terahertz Waves **41**, 470–489 (2020)
4. Specim: Whitepaper. Recent developments in hyperspectral imaging in industrial applications. http://www.specim.com. Accessed 25 Feb 2025
5. Linowska, B., Garbacz, P.: Hyperspectral imaging system for food safety inspection. In: Szewczyk, R., Zieliński, C., Kaliczyńska, M., Bučinskas, V. (eds.) Automation 2023: Key Challenges in Automation, Robotics and Measurement Techniques. AUTOMATION 2023, LNCS, vol. 630, pp. 204—211. Springer, Cham (2023). https://doi.org/10.10007/1234567890
6. Cheng, Y., Deng, Y., Cao, J., Xiong, X., Bai, L., Li, Z.: Dual-band multi-wave and hybrid imaging techniques: a new direction for nondestructive testing and structural health monitoring. Sensors **13**, 16146–16190 (2013)

7. Marciniak, T., Lutowski, Z., Bujnowski, S., Boroński, D., Czajka, P.: Dual-band experimental system for subsurface cracks testing. Mater. Sci. Forum **726**, 222–225 (2012)
8. Mitchell, H.B.: Multi-Sensor Data Fusion, 2nd edn. Springer, Heidelberg (2007). https://doi.org/10.1007/978-3-540-30301-5_26
9. Lahat, D., Adali, T., Jutten, C.: Multimodal data fusion: an overview of methods, challenges, and prospects. Proc. IEEE **103**(9), 1449–1477 (2015)
10. Zhang, Y., Lee, H.-J.: A target-aware fusion framework for infrared and visible images. IEEE Access **11**, 33666–33681 (2023)
11. Han, M., et al.: Boosting target-level infrared and visible image fusion with regional information coordination. Inf. Fus. **92**, 268–288 (2023)
12. Ma, J., Ma, Y., Li, Ch.: Infrared and visible image fusion methods and applications: a survey. Inf. Fus. **45**, 153–178 (2019)
13. Sannen, D., Brussel, H.: A multilevel information fusion approach for visual quality inspection. Inf. Fus. **13**(1), 48–59 (2012)
14. Cao, Y., et al.: Depth and thermal sensor fusion to enhance 3D thermographic reconstruction. Opt. Express **13**, 8179–8193 (2018)
15. Pietrzkiewicz P.: Image analysis techniques for industrial inspection systems. https://www.adaptive-vision.com/download/mvg.pdf. Accessed 25 Feb 2025
16. Dandrifosse, S., Carlier, A., Dumont, B., Mercatoris, B.: Sannen, D., Brussel, H.:Registration and fusion of close-range multimodal wheat images in field conditions. Remote Sens. **13**(7), 1380–8193 (2021)
17. Zitová, B., Flusser, J.: Image registration methods: a survey. Image Vis. Comput. **21**(11), 977–1000 (2003)
18. ACERTA: Whitepaper. The future of quality in manufacturing. https://acerta.ai/predictive-quality-whitepaper/. Accessed 25 Feb 2025
19. Shahsavarani, S., Lopez, F., Ibarra-Castanedo, C., Maldague, X.: Multi-modal image registration for image fusion enhancement in infrastructure inspection. Sensors **24**, 3994 (2024)
20. Zebra: Brochure aurora^{TM} deep learning. www.zebra.com/. Accessed 2 May 2025

Cascade of One-Class Classifier Ensemble and Dynamic Naive Bayes Classifier Applied to the Myoelectric-Based Upper Limb Prosthesis Control with Contaminated Channels Detection

Pawel Trajdos(✉) and Marek Kurzynski

Wroclaw University of Science and Technology, 27 Wybrzeze Stanislawa Wyspianskiego st., 50-370 Wroclaw, Poland
{pawel.trajdos,marek.kurzynski}@pwr.edu.pl

Abstract. Modern upper limb bioprostheses are typically controlled by sEMG signals using a pattern recognition scheme in the control process. Unfortunately, the sEMG signal is very susceptible to contamination that deteriorates the quality of the control system and reduces the usefulness of the prosthesis in the patient's everyday life. In the paper, the authors propose a new recognition system intended for sEMG-based control of the hand prosthesis with detection of contaminated sEMG signals. The originality of the proposed solution lies in the co-operation of two recognition systems working in a cascade structure: (1) an ensemble of one-class classifiers used to recognise contaminated signals and (2) a naive Bayes classifier (NBC) which recognises the patient's intentions using the information about contaminations produced by the ensemble. Although in the proposed approach, the NBC model is changed dynamically, due to the multiplicative form of the classification functions, training can be performed in a one-shot procedure. Experimental studies were conducted using real sEMG signals. The results obtained confirm the hypothesis that the use of the one-class classifier ensemble and the dynamic NBC model leads to improved classification quality.

Keywords: myoelectric hand prosthesis control · contaminated sEMG signal · one-class classifier ensemble · naive Bayes classifier

1 Introduction

In the 1950s, the concept of controlling an active, externally powered hand prosthesis using an sEMG signal was introduced. This initiated the intensive development of advanced myoelectrically controlled anthropomorphic upper limb prostheses, which is still ongoing [5]. Modern bioprostheses are typically controlled by sEMG signals from residual limb muscles using a pattern recognition scheme in the control process. In this approach, the patient's intention of moving the prosthesis (manipulation or grip) denotes the object being recognised, the recorded

P. Trajdos and R. Burduk (Eds.): CORES 2025, LNNS 1557, pp. 30–42, 2025.
https://doi.org/10.1007/978-3-032-01773-4_4

sEMG signals constitute the object representation, and the type of movement of the prosthesis is the class label [18].

This means that the correct operation of the control system, which determines the comfort of using the prosthesis in patient's everyday life, depends exclusively on the quality of the overall system for recognizing the patient's intention. The essence of the recognition system is a sequence of actions performed on the electromyographic signal by a human (patient), a measuring system, and a classifier. Each stage of this sequence may be a source of errors that deteriorate the quality of the control system [7].

The topic of this paper is to address the recognition of contaminated sEMG signals within the context of myoelectric control for upper limb prosthetics in order to improve the quality of the control process.

Due to its low power and unfavourable recording conditions, the sEMG signal is very susceptible to contaminations, which can be divided into three main categories [13]: 1. noise (thermal and flicker noise, amplifier saturation, analogue to digital signal clipping, quantisation noise), 2. external interference (power line, ECG, crosstalk), 3. artifacts (measurement artifacts, baseline wander, motion artifact, electrode shifts, changes of electrode/skin impedance (sweating).)

Generally, two types of methods are proposed in the literature to mitigate the adverse effects of sEMG signal contamination. The first type of methods is aimed at directly improving the quality of the recorded sEMG signal, which can be achieved by better hardware [6] or by software filters tailored to characteristics of different contamination of sEMG signal [4].

The second category includes methods based on the artificial intelligence paradigm, in which information about sEMG signal contamination is extracted, and then this information is used in model of a decision-making system to improve the quality of the control process.

Particularly noteworthy are the methods with a common concept based on the use of two multiclassifier recognition systems. The first system is an ensemble of one-class classifiers that assess the degree of contamination of individual sEMG signals (channels). The purpose of the second system is to classify the movement of the prosthesis, and its recognition procedure is supported by the results from a one-class classifier ensemble. The advantage of this solution is the completeness of the patient's intention recognition with detection of contaminated channels and independence of the type of contamination, because one-class classifiers are trained exclusively on clean signals. This concept is quite richly represented in the literature [8,9], including the authors' earlier works [17,19] . The proposed solutions differ in the decision scheme of the one-class classifiers (the method of assessing signal contamination), the number of channels (signals) used in the base classifiers of the second system and the classifiers combining procedure.

In the paper, the authors develop a new recognition system intended for sEMG-based control of the upper limb prosthesis with detection of contaminated sEMG signals, which also fits into the concept described above. The originality of the proposed solution lies in the co-operation of an ensemble of single-class classifiers and a naive Bayes classifier working in a cascade structure. In the first step, the ensemble of one-class classifiers is typically used to recognise contaminated channels, returning information about the level of contamination for each channel. In the second step, the naive Bayes classifier recognises the patient's intentions using the information produced by the ensemble. This means that the model of the naive Bayes classifier is dynamically changing, as the information about contamination may differ between objects. The choice of the naive Bayes classifier for recognition of patient's intention was deliberate and justified by the multiplicative form of its classification functions, which allowed for training the classifier in a one-shot procedure.

The main goal of the paper and the conducted experimental study is to answer the following research questions: – **RQ1** Does the proposed method perform better than a simple classifier trained on the entire feature space? – **RQ2** Does the proposed method perform better than the alternative approaches presented in the literature? – **RQ3** How do the investigated methods perform under different values of signal-to-noise ratio?

The paper is organised as follows. Section 2 presents the models of two classification systems in a cascade structure and the method for their cooperation. Section 3 describes the experimental setup. The results of experimental studies with discussion are presented in Sect. 4. Section 5 concludes the paper.

2 Methods

Let us consider a bioprosthetic hand control system based on a multi-channel sEMG signal recognition scheme. Let

$$\mathcal{C} = \{C_1, C_2, \dots, C_L\} \tag{1}$$

Denotes the set of sEMG signals recorded from the L sensors (channels) located on the patient's forearm stump. The signals (1) represent the patient's intention to perform a specific movement of the prosthesis. The activation of residual stump muscles, which are the source of sEMG signals, is based on the phantom movement of the amputated limb that the user is able to voluntarily control [1].

Let $x_l = (x_l^{(1)}, x_l^{(2)}, \dots, x_l^{(d_l)})$ be a feature vector belonging to the feature space $\mathcal{X}_l$, which was extracted from the sEMG signal C_l ($l = 1, 2, \dots, L$). Consequently, $x = (x_1, x_2, \dots, x_L) \in \mathcal{X}$ denotes the whole feature vector describing the patient's intention, and $\mathcal{X} = \mathcal{X}_1 \times \mathcal{X}_2 \times \dots \times \mathcal{X}_L$.

Movements executed by the prosthesis are labeled by consecutive natural numbers, forming a set of classes for the recognition task:

$$\mathcal{M} = \{1, 2, \dots, M\}. \tag{2}$$

The number of classes M can vary among patients, as it is related to their ability to activate the muscles of the stump.

We assume that sEMG signals (1) can be contaminated during the acquisition process. Contaminants can randomly occur in various sEMG signal channels, making it impossible to predict in advance the number of affected channels and their identities.

Signal contaminants can considerably reduce the classification quality of a recognition system. Therefore, our objective is to develop a recognition system capable of identifying signal contamination and leveraging these data to enhance the classification quality of contaminated samples.

The proposed method is based on two cooperating classification systems in a cascade structure. The first system is an ensemble of one-class classifiers whose task is to recognise contaminated sEMG signals.

The second system is a Naive Bayes classifier (NB) working in dynamic mode and using the information provided by the previous one-class ensemble system.

The following subsections elaborate on the specifics of both classifier systems.

2.1 Ensemble of One-Class Classifiers

An ensemble of one-class classifiers is denoted as:.

$$\Phi = \{\phi_1, \phi_2, \dots, \phi_L\}. \tag{3}$$

The base classifier ϕ_l of the system (3) uses the feature vector x_l extracted from signal C_l and produces a prediction

$$\phi_l(x_l) = r_l, \ l = 1, 2, \dots, L, \tag{4}$$

that evaluates the contamination of signal C_l. The interpretation of the prediction (4) depends on the decision scheme adopted by the system (3).

Crisp decision scheme, in which $r_l \in \{0, 1\}$. Zero indicates that x_l is an outlier in the feature space $\mathcal{X}_l$ and the corresponding signal C_l is recognised as contaminated with noise. Consequently, $r_l = 1$ indicates that x_l is recognised as a target class object, i.e., the corresponding signal C_l is considered free of contaminants.

Soft decision scheme, in which $r_l \in [0, 1]$. The closer to 0 r_l, the level of contaminants in the registered signal is greater.

Moreover, let $r = (r_1, r_2, \dots, r_L) \in \mathcal{R}$ be a vector of predictions produced by the ensemble Φ, where $\mathcal{R} = \{0, 1\}^L$ $(= [0, 1]^L)$ for the crisp (soft) scheme, respectively.

Each base classifier (4) is trained using a channel-specific training set that contains only objects of the target class without the presence of outliers (contaminated signals). This is the usual way to build one-class classifiers [16].

2.2 Naive Bayes Classifier in Dynamic Mode

The second classifier in the cascade is designed to recognise the class of prosthesis movement from the set (2) according to the patient's intention:

$$\psi(x, r): \ \mathcal{X} \times \mathcal{R} \to \mathcal{M}. \tag{5}$$

The classifier (5), in addition to the typical dependence on the feature vector x, uses additional information on the contamination of individual sEMG signals C_l contained in the prediction vector r generated by the one-class system (3).

Now the key question arises: How can the prediction vector r in the classifier (5) be used to achieve improvement in classification quality? In the proposed method, this is done by eliminating features originating from contaminated signals (for the crisp scheme) or by weighting the influence of features depending on the degree of channel contamination (for the soft scheme).

To demonstrate this mechanism in detail, let us assume the canonical classifier model which means that for a given feature vector x and for prediction vector r the classifier produces a vector of supports (values of the classifying functions) for each class:

$$d(x, r) = [d_1(x, r), d_2(x, r), \dots, d_M(x, r)], \tag{6}$$

where $d_j(x, r) \in [0, 1]$ and $\sum_j d_j(x, r) = 1$. The class with the highest support is the final recognition result of the classifier ψ.

For this classification model, the mechanism of influence of the prediction vector r on classifier performance is quite intuitive. Namely, it is necessary to increase (decrease) the value of supports depending on the purity (contamination) of the signals from which the features are extracted. Unfortunately, this approach has a significant drawback, as the signal contamination phenomenon is dynamic in nature, which means that the configuration of contaminated channels will probably be different for each recognised object. This requires dynamic (for each object) training of the classifier ψ. Such a solution is out of the question because the bioprosthesis control system is a real-time system (the time from the beginning of sEMG signal recording to prosthesis movement should not be longer than 200 ms).

A classifier with a Naive Bayes (NB) model can be helpful here, as it eliminates the problem of dynamic learning due to the assumption of feature independence. The NB classifier is based on a probabilistic model in which the supports are equal to the posterior probabilities for the individual classes $d_j(x, r) = p(j|x; r)$, $j \in \mathcal{M}$, where r denotes the parameter of these probabilities. Using Bayes' formula we can express the posterior probabilities in the form:

$$p(j|x; r) = c \cdot p_j \cdot P(x|j; r), \quad j \in \mathcal{M}, \tag{7}$$

where c denotes normalizing factor, p_j is a prior probability of the j-th class and $P(\cdot)$ denotes probability distribution of feature vector x given that object is from the jth class.

Assuming conditional independence of features, we have

$$P(x|j; r) = \prod_{l=1}^{L} \prod_{i=1}^{d_l} P(x_l^{(i)}|j; r) \tag{8}$$

where $P(\cdot)$ denotes the conditional probability distribution (probability or density function for a discrete or continuous feature, respectively). In order to obtain a formula for the supports $d_j(x, r)$ from (7) and (8) we still need to take into account the dependence on the prediction r. According to the mechanism proposed and presented previously, the value of $P(x_l^{(i)}|j)$ will decrease depending on the level of contamination r_l of the channel (signal) C_l. This effect will be obtained for crisp and soft decision schemes by raising the appropriate probability to the power r_l. This leads to the following final form of the classification functions of the naive Bayes algorithm ψ, and at the same time the entire two-stage patient's intention recognition system with identification of contaminated channels:

$$d_j(x, r) = c \cdot p_j \cdot \prod_{l=1}^{L} \left[\prod_{i=1}^{d_l} P(x_l^{(i)}|j) \right]^{r_l} \quad j \in \mathcal{M}. \tag{9}$$

This formula defines an attribute-weighted Naive Bayes classifier [12].

Note that by assuming conditional independence of features and defining decision functions as the product of probability distributions for single features, we do not need to train the classifier dynamically for each subsequent test object. All that needs to be done is to determine all elementary probability distributions $P(x_l^{(i)}|j)$ (their number is $\sum_l d_l$) in a one-shot training process, which in this case consists of estimating these distributions from the training set treated as a random sample. Finally, the functions (9) are dynamically determined and compared for different classes to identify the class with the highest support value.

3 Experimental Setup

The experimental study is conducted to answer the research questions posed in Sect. 1. To do so, we compared the following methods: – **B**: A Naive Bayes classifier trained on the data coming from all available EMG channels. – **EC**: Error-correcting output codes ensemble trained on the data from all available EMG channels [15]. We considered the following values of *code_size* parameter $\{2, 3, 4, 5, 6\}$ The parameters are tuned with a simple search approach, using four-fold cross-validation. – **NBH**: Naive Bayes classifier with attributes selected in a crisp way. – **NBS**: Soft Naive Bayes approach.

In this study, we employed the One-class **SVM** classifier with RBF kernel [8]. This is because in our previous studies this classifier achieved the best classification performance in the task of identifying noise-contaminated sEMG channels [17]. The classifier has a default method to adjust the parameter γ. The ν parameter of this classifier is tuned. The following values of ν are considered $\{0.1, 0.2, \ldots, 1.0\}$.

To determine the best values for the parameters listed above, the following method is utilised. A four-fold, cross-validation approach is employed to create the training and validation sets. The validation segment is subsequently enhanced by incorporating artificial examples labelled as noise class. Artificial examples are generated using a uniform distribution in the classifier-specific input space $\mathcal{X}_l$. This is because the main motivation for using one-class classifiers instead of binary classifiers is to avoid making any assumptions about the outlier distribution. When the best value of the parameters is found, the final one-class classifier is trained using the entire training set. For parameter tuning, we use the balanced accuracy score as a quality measure.

The following versions of the Naive Bayes classifier are considered in the experiments: – Naive Bayes classifier using a Gaussian distribution estimator (NBG). – Naive Bayes classifier using a mixture of Gaussian distributions (NBGMT). The number of distributions is tuned using four-fold cross-validation and balanced accuracy as the quality criterion. The following numbers of Gaussian distributions are considered: $\{1, 3, 5, 7\}$. The seeds for Gaussian distributions are set using the K-means++ approach [3].

We used the one-class and multiclass classifiers implemented in the scikit library [14]. Since the soft outputs of the one-class classifiers used are incompatible with the model presented in (9), we applied the logistic regression model [20] to make them compatible with the model. Unless otherwise specified, the classifier parameters are set to their default values.

The signals used in these experiments come from the Web repository[1]. The demographic information details for each amputee and the details of the dataset are presented in [2]. To unify the signals coming from different subjects, we used sEMG signals from the first 8 channels. We selected the signals associated with a low force level. In separating individual objects for particular classes, we used a non-overlapped segmentation scheme with a segment length of 500 ms. This resulted in 70 to 310 objects for each class.

To simulate real-world EMG and MMG signal contaminations, the following noise generation techniques are used [4,7] – Simulation of power grid whose frequency varies from 48 to 52 Hz. The amplitude of the inserted noise depends on the SNR. – Signal attenuation that simulates the sensor losing contact with the skin. The attenuation level depends on the selected SNR. – Gaussian noise. It simulates the general noises that may appear in the signal acquisition circuit. – Simulation of non-linear amplifier characteristics for signals of high amplitude. This is done by non-linear clipping the peaks of the signal. The clipping level depends on the selected SNR. – Baseline wandering. This is simulated by injecting low-frequency (0.5 to 1.5 Hz) sinusoidal noise. The amplitude of the signal depends on the selected SNR.

In experimental studies, we consider the following SNR levels $\{0, 1, 2, \ldots, 6, 10, 12\}$.

The training and testing sets are obtained by a ten-fold cross-validation repeated 3 times. Each example from the testing dataset is then randomly contaminated with one of the above-mentioned noise types with a selected SNR level. The original testing set is finally extended using samples with artificial noise.

Feature vectors were created from raw EMG signals using the discrete wavelet transform technique. The *db6* wavelet and three levels of decomposition were

[1] https://www.rami-khushaba.com/.

used. The following functions were calculated for the transformation coefficients [11]: MAV, SSC.

To asses the classification quality, we employed the following quality criteria: – Balanced Accuracy (BAC), – Cohen's kappa coefficient, – Micro-averaged F_1 measure.

The statistical significance of the results obtained was evaluated using the pairwise Wilcoxon signed rank test. Family-wise errors were using Holm's procedure. The significance level was set to $\alpha = 0.05$ [10]. For some analysis, the average rank approach is also used. The experimental code is provided in[2].

4 Results and Discussion

The average rank plots are presented in Figs. 1 and 2. The exact average rank values and the results of the statistical tests are shown in Tables 1 and 2. Each table is divided into quality-criterion-specific sections. In each section, the column names are related to the investigated methods. In each row, the average ranks for a given SNR value are presented. The subscript under the average rank value presents the outcome of the statistical tests, with the subscript containing a comma-separated list of numbers or pause '–'. The list contains column numbers of methods that are significantly better than a given method. Pause means that the method is not significantly better than any other.

The results presented for both base classifiers **NBG** and **NBGMT** and all quality measures are pretty consistent. Let us describe the main thrends presented in the data. First of all, according to the average ranks, the **EC** ensemble is the worst method in this comparison. The results show that in the scenario considered, the decomposition based on the error-correcting output codes does not perform well in combination with the Naive Bayes classifier. The reason may be the fact that the method decomposes the multi-class classification problem ino multiple binary problems and then combines the results. The ensemble creation method is unaware that some data may be noisy.

The best performing methods for a wide range of SNR values are **NBS** and **NBH** methods. That is surely because the method works together with one-class-svm based outlier detectors. Consequently, they are aware of the presence of noise in the input data. The differences between them and the base Naive Bayes classifier (**B**) are greater when the SNR value is lower. That is, the dynamic Naive Bayes classifiers **NBH** and **NBS** tend to be better when the amount of noise introduced in the data is greater. This is because when the amount of noise is high, the base, Naive Bayes classifier **B** is no longer robust against noise and elimination of affected channels causes the classification results to be better. When the amount of noise is lower (higher SNR values), the base Naive Bayes classifier seems to be robust to the noise present in the data and its results become comparable to **NBS** and **NBH** methods. This is especially visible in the case of weaker **NBG** base classifier. For **NBG** approach and SRN = 12, **B**

[2] https://github.com/ptrajdos/CORES_2025.git.

method is significantly better than the dynamic approaches **NBH** and **NGS**. This may be due to the fact that, for SNR=12 the amount of noise is low enough for **B** to deal with it efficiently. In this scenario, any misclassification at the first level of the ensemble (one-class SVM classifier) has a significant negative impact on the overall classification accuracy. However, the phenomenon is not observed for the **NBGMT** approach.

When we compare the dynamic Naive Bayes approaches, some significant differences can also be observed. First, the **NBS** method tends to outperform the **NBH** method in most cases. In other cases the methods are comparable. There are no cases when **NBH** is significantly better than **NBS**. This means that weighting the attributes taking the amount of noise present in the related sEMG channel allows the classifier to perform better under noisy conditions. This is because the channel that contains a low amount of noise would give the method some information related to the object that is currently classified. The reason is related to the robustness of the base classifier to some degree of noise present in the data. Due to this robustness, completely eliminating the noisy channels is not a good strategy.

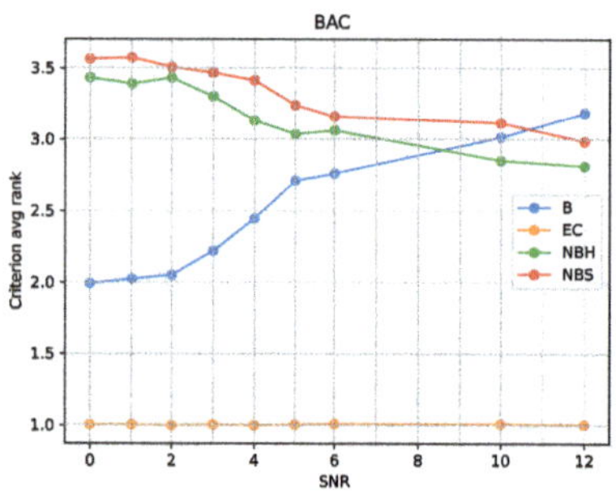

(a) Balanced accuracy criterion

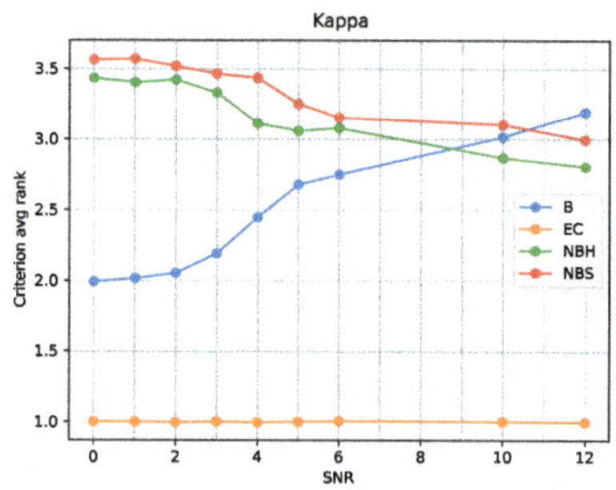

(b) Kappa criterion

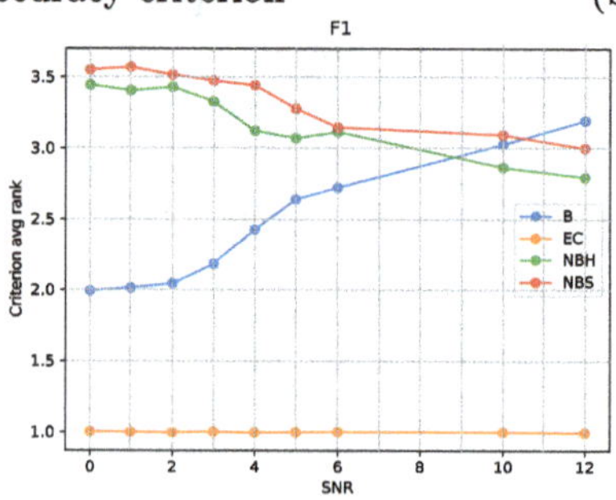

(c) F_1 criterion

Fig. 1. Classification quality for NBG classifier – average ranks.

To reiterate, the main research questions posed in Sect. 1 are as follows: – **RQ1** Yes, the proposed method outperforms the simple classifier in most cases.

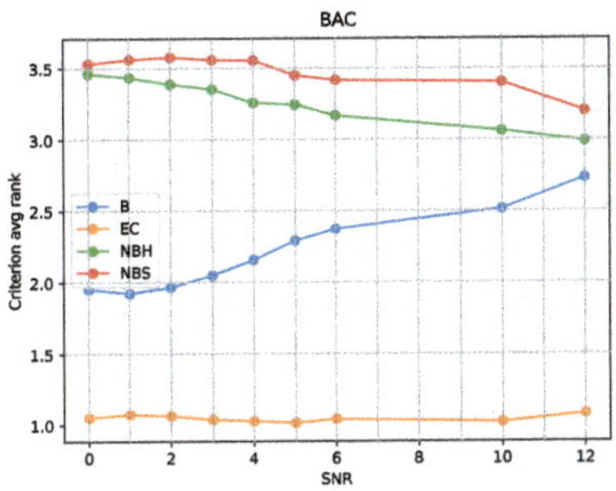

(a) Balanced accuracy criterion

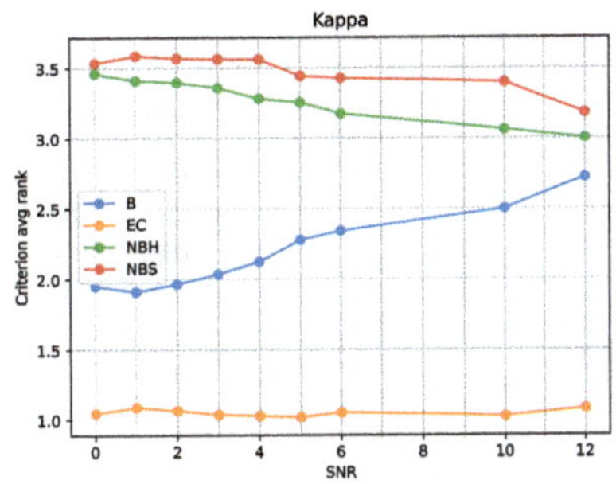

(b) Kappa criterion

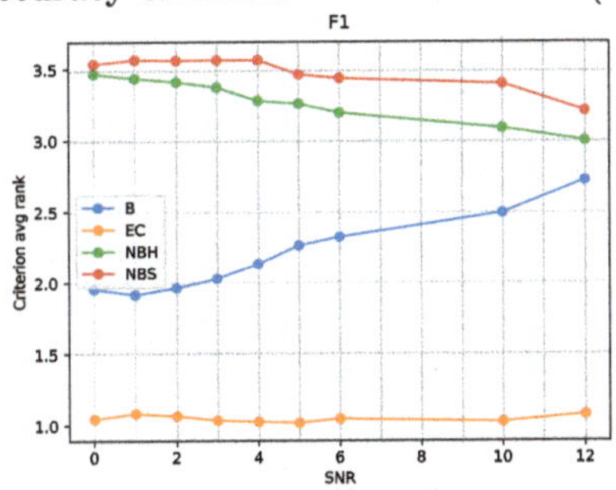

(c) F_1 criterion

Fig. 2. Classification quality for NBGMT classifier – average ranks.

Table 1. Classification quality for NBG classifier, statistical tests.

	BAC				Kappa				F1			
	B	EC	NBH	NBS	B	EC	NBH	NBS	B	EC	NBH	NBS
SNR:12	3.190	1.005	2.817	2.988	3.194	1.000	2.808	2.998	3.197	1.000	2.799	3.005
	2,3,4	–	2	2,3	2,3,4	–	2	2,3	2,3,4	–	2	2,3
SNR:10	3.019	1.009	2.854	3.118	3.019	1.005	2.870	3.106	3.030	1.005	2.868	3.097
	2	–	2	2	2	–	2	2	2	–	2	2
SNR:6	2.764	1.009	3.065	3.162	2.752	1.009	3.081	3.157	2.727	1.005	3.118	3.150
	2	–	1,2	1,2	2	–	1,2	1,2	2	–	1,2	1,2
SNR:5	2.713	1.005	3.039	3.243	2.681	1.005	3.060	3.255	2.644	1.002	3.074	3.280
	2	–	1,2	1,2,3	2	–	1,2	1,2,3	2	–	1,2	1,2,3
SNR:4	2.449	1.000	3.134	3.417	2.449	1.000	3.113	3.438	2.431	1.000	3.125	3.444
	2	–	1,2	1,2,3	2	–	1,2	1,2,3	2	–	1,2	1,2,3
SNR:3	2.222	1.005	3.303	3.470	2.194	1.005	3.333	3.468	2.188	1.005	3.331	3.477
	2	–	1,2	1,2,3	2	–	1,2	1,2,3	2	–	1,2	1,2,3
SNR:2	2.056	1.000	3.435	3.509	2.056	1.000	3.424	3.521	2.049	1.000	3.433	3.519
	2	–	1,2	1,2	2	–	1,2	1,2	2	–	1,2	1,2
SNR:1	2.028	1.005	3.394	3.574	2.019	1.005	3.405	3.572	2.019	1.002	3.407	3.572
	2	–	1,2	1,2,3	2	–	1,2	1,2,3	2	–	1,2	1,2,3
SNR:0	1.995	1.005	3.435	3.565	1.995	1.005	3.435	3.565	1.995	1.005	3.447	3.553
	2	–	1,2	1,2,3	2	–	1,2	1,2,3	2	–	1,2	1,2,3

Table 2. Classification quality for NBGMT classifier, statistical tests.

	BAC				Kappa				F1			
	B	EC	NBH	NBS	B	EC	NBH	NBS	B	EC	NBH	NBS
SNR:12	2.731	1.079	2.991	3.199	2.725	1.083	3.007	3.185	2.720	1.076	2.998	3.206
	2	–	1,2	1,2,3	2	–	1,2	1,2,3	2	–	1,2	1,2,3
SNR:10	2.514	1.021	3.062	3.403	2.502	1.028	3.067	3.403	2.493	1.025	3.086	3.396
	2	–	1,2	1,2,3	2	–	1,2	1,2,3	2	–	1,2	1,2,3
SNR:6	2.373	1.042	3.169	3.417	2.343	1.051	3.178	3.428	2.324	1.046	3.194	3.435
	2	–	1,2	1,2,3	2	–	1,2	1,2,3	2	–	1,2	1,2,3
SNR:5	2.292	1.014	3.245	3.449	2.280	1.019	3.257	3.444	2.264	1.019	3.257	3.461
	2	–	1,2	1,2,3	2	–	1,2	1,2,3	2	–	1,2	1,2,3
SNR:4	2.157	1.028	3.259	3.556	2.125	1.028	3.285	3.562	2.134	1.025	3.278	3.562
	2	–	1,2	1,2,3	2	–	1,2	1,2,3	2	–	1,2	1,2,3
SNR:3	2.051	1.037	3.354	3.558	2.037	1.037	3.361	3.565	2.030	1.035	3.373	3.562
	2	–	1,2	1,2,3	2	–	1,2	1,2,3	2	–	1,2	1,2,3
SNR:2	1.968	1.065	3.389	3.579	1.968	1.065	3.398	3.569	1.965	1.065	3.410	3.560
	2	–	1,2	1,2,3	2	–	1,2	1,2,3	2	–	1,2	1,2,3
SNR:1	1.926	1.074	3.438	3.562	1.912	1.088	3.412	3.588	1.917	1.083	3.435	3.565
	2	–	1,2	1,2,3	2	–	1,2	1,2	2	–	1,2	1,2
SNR:0	1.954	1.051	3.463	3.532	1.954	1.046	3.463	3.537	1.956	1.044	3.465	3.535
	2	–	1,2	1,2	2	–	1,2	1,2	2	–	1,2	1,2

The differences are more pronounced when the SNR value is lower. – **RQ2** Yes, the proposed method performs better than the alternative approaches presented in the literature. The experimental results demonstrate that the dynamic Naive Bayes classifiers (**NBH** and **NBS**) outperform the error-correcting output codes ensemble (**EC**) and the base Naive Bayes classifier (**B**) in most cases. The differences are statistically significant, especially for lower SNR values, where the dynamic classifiers effectively handle noisy data. – **RQ3** The investigated methods perform differently under varying SNR values. The dynamic Naive Bayes classifiers (**NBH** and **NBS**) show better performance at lower SNR values, where noise contamination is higher. The **NBS** method, which uses a soft weighting scheme, generally outperforms the crisp approach (**NBH**), as it can leverage partially contaminated channels more effectively. At higher SNR values, the performance gap between the dynamic classifiers and the base Naive Bayes classifier (**B**) narrows, as the impact of noise diminishes.

5 Conclusions

The objective of this paper is to develop a multiclass recognition system capable of identifying signal contamination and leveraging these contamination-related data to improve the classification of contaminated samples. The proposed method is based on two cooperating classification systems in a cascade structure. The first system is an ensemble of one-class classifiers whose task is to recognise contaminated sEMG signals. The second system is a Naive Bayes classifier (NB) working in dynamic mode and using the information provided by the previous one-class ensemble system. This is done either by eliminating features originating from contaminated signals or by weighting the influence of features depending on the degree of channel contamination (for the soft scheme). The training and validation sets are subsequently enhanced by incorporating artificial examples labelled as noise class. The greatest advantage of the dynamic Naive Bayes classifier is that it does not need any retraining to order to change classification model by removing or changing the weights of attributes. In addition, the experimental study shows that using the information from the one-class classifiers allows the entire ensemble to improve the classification quality significantly compared to the reference methods. The experiments also show that the sof version of the dynamic Naive Bayec classifier performs significantly better than the crisp version. The results are very promising, and the concept of using a dynamic Naive Bayes classifier in the context of classification of contaminated sEMG signals should be further explored.

References

1. Akbulut, A., Gungor, F., Tarakci, E., Aydin, M.A., Zaim, A.H., Catal, C.: Identification of phantom movements with an ensemble learning approach. Comput. Biol. Med. **150**, 106132 (2022). https://doi.org/10.1016/j.compbiomed.2022.106132
2. Al-Timemy, A.H., Khushaba, R.N., Bugmann, G., Escudero, J.: Improving the performance against force variation of EMG controlled multifunctional upper-limb prostheses for transradial amputees. IEEE Trans. Neural Syst. Rehabil. Eng. **24**(6), 650–661 (2016). https://doi.org/10.1109/tnsre.2015.2445634
3. Arthur, D., Vassilvitskii, S.: k-means++: The advantages of careful seeding. In: Proceedings of the Eighteenth Annual ACM-SIAM Symposium on Discrete Algorithms, pp. 1027–1035. Society for Industrial and Applied Mathematics (2007)
4. Boyer, M., Bouyer, L., Roy, J.S., Campeau-Lecours, A.: Reducing noise, artifacts and interference in single-channel EMG signals: a review. Sensors **23**(6), 2927 (2023). https://doi.org/10.3390/s23062927
5. Chen, Z., Min, H., Wang, D., Xia, Z., Sun, F., Fang, B.: A review of myoelectric control for prosthetic hand manipulation. Biomimetics **8**(3), 328 (2023). https://doi.org/10.3390/biomimetics8030328
6. Farago, E., Chan, A.: Detection and reconstruction of poor-quality channels in high-density EMG array measurements. Sensors **23**(10), 4759 (2023). https://doi.org/10.3390/s23104759
7. Farago, E., MacIsaac, D., Suk, M., Chan, A.: A review of techniques for surface electromyography signal quality analysis. IEEE Rev. Biomed. Eng. **16**, 472–486 (2023). https://doi.org/10.1109/rbme.2022.3164797
8. Fraser, G.D., Chan, A., Green, J.R., MacIsaac, D.T.: Automated biosignal quality analysis for electromyography using a one-class support vector machine. IEEE Trans. Instrum. Meas. **63**(12), 2919–2930 (2014). https://doi.org/10.1109/tim.2014.2317296
9. ichiro Furukawa, J., Noda, T., Teramae, T., Morimoto, J.: Fault tolerant approach for biosignal-based robot control. Adv. Robot. **29**(7), 505–514 (2015). https://doi.org/10.1080/01691864.2014.996603
10. Garcia, S., Herrera, F.: An extension on"statistical comparisons of classifiers over multiple data sets"for all pairwise comparisons. J. Mach. Learn. Res. **9**, 2677–2694 (2008)

11. Junior, J.J.A.M., Freitas, M.L., Siqueira, H.V., Lazzaretti, A.E., Pichorim, S.F., Stevan, S.L.: Feature selection and dimensionality reduction: an extensive comparison in hand gesture classification by sEMG in eight channels armband approach. Biomed. Sig. Process. Control **59**, 101920 (2020). https://doi.org/10.1016/j.bspc.2020.101920
12. Kalra, M., Kumar, V., Kaur, M., Ahmed Idris, S., Öztürk, c., Alshazly, H.: Attribute weighted naïve bayes classifier. Comput. Mater. Continua **71**(1), 1945–1957 (2022). https://doi.org/10.32604/cmc.2022.022011
13. Kyranou, I., Vijayakumar, S., Erden, M.S.: Causes of performance degradation in non-invasive electromyographic pattern recognition in upper limb prostheses. Front. Neurorobot. **12** (2018). https://doi.org/10.3389/fnbot.2018.00058
14. Pedregosa, F., et al.: Scikit-learn: machine learning in Python. J. Mach. Learn. Res. **12**, 2825–2830 (2011)
15. Sarabia, P., Araujo, A., Sarabia, L.A., de la Cruz Ortiz, M.: Electromyography gesture model classifier for fault-tolerant-embedded devices by means of partial least square class modelling error correcting output codes (PLS-ECOC). Algorithms **16**(3), 149 (2023). https://doi.org/10.3390/a16030149
16. Seliya, N., Abdollah Zadeh, A., Khoshgoftaar, T.M.: A literature review on one-class classification and its potential applications in big data. J. Big Data **8**(1), 1–31 (2021). https://doi.org/10.1186/s40537-021-00514-x
17. Trajdos, P.: A noise-tolerant dual multi-classifier system with fuzzy model applied to the semg-based control of a bionic upper limb prosthesis. Biomed. Sig. Process. Control **104**, 107441 (2025). https://doi.org/10.1016/j.bspc.2024.107441
18. Trajdos, P., Kurzynski, M.: Application of context-dependent interpretation of biosignals recognition to control a bionic multifunctional hand prosthesis. Biocybern. Biomed. Eng. **44**(1), 161–182 (2024). https://doi.org/10.1016/j.bbe.2024.01.001
19. Trajdos, P., Kurzynski, M.: A dual ensemble classifier used to recognise contaminated multi-channel EMG and MMG signals in the control of upper limb bioprosthesis. In: Zyblewski, P., Grana, M., Pawel, K., Minku L. (eds.) Proceedings of The Workshop on Classifier Learning from Difficult Data, Proceedings of Machine Learning Research, vol. 263, pp. 1–8. PMLR (2024)
20. Yu, H.F., Huang, F.L., Lin, C.J.: Dual coordinate descent methods for logistic regression and maximum entropy models. Mach. Learn. **85**(1–2), 41–75 (2010). https://doi.org/10.1007/s10994-010-5221-8

Machine Learning to Predict Be-7 Concentration in Europe

Ander Nafarrate[1(✉)], Raquel Idoeta[1], Alain Ulazia[2], Gabriel Ibarra[1], and Susana Petisco-Ferrero[1]

[1] Energy Engineering Department, University of the Basque Country, UPV/EHU, Bilbao, Spain
ander.nafarrate@ehu.eus

[2] Energy Engineering Department, University of the Basque Country, UPV/EHU, Eibar, Spain

Abstract. Observed airborne ^{7}Be activity concentration dependence on atmospheric variables is studied with the aim of characterising the behaviour of this radionuclide. Time series corresponding to 24 sites covering a significant range of longitudes and latitudes across Europe are modelled using the random forest algorithm. The Kolmogorov-Zurbenko filter is applied to each time series in order to separate the clean main signal from its noisy fraction. The clean fractions are then used as the target of the machine learning models, which are fed with surface and upper atmosphere variables from the ERA5 reanalysis and observed daily sunspot number time series. The validation of the models is addressed in terms of coefficient of determination (R^2) and Mean Absolute Error; a process in which most of the generated models appear to be representative enough to allow a significant feature importance analysis. The relevance of the features is divided into variables related to the ^{7}Be cosmogenic generation mechanism, its atmospheric precipitation mechanism and the autocorrelation of the ^{7}Be time series.

Keywords: ^{7}Be · Random Forest · Machine learning · KZ filter

1 Introduction

Due to their distribution characteristics in the atmosphere, radionuclides of cosmogenic origin have utility as tools for enhancing the comprehension of climate change and atmospheric processes, including air and aerosol transport. ^{7}Be ($T_{1/2}$ = 53.22 d) is a radionuclide produced in the stratosphere and the upper troposphere primarily by the spallation reactions of high-energy protons with nitrogen or oxygen atoms. Following its production, ^{7}Be nuclides attach to atmospheric aerosols and follow their transport through the atmosphere, from where they are removed by dry or wet deposition and by radioactive decay [3,7]. ^{7}Be activity concentration in aerosols can be easily measured. Its value is dependent on the characteristics of its production mechanism, namely solar activity, latitude,

P. Trajdos and R. Burduk (Eds.): CORES 2025, LNNS 1557, pp. 43–53, 2025.
https://doi.org/10.1007/978-3-032-01773-4_5

altitude, and also by meteorological dynamics [10]. In this regard, ^{7}Be can be considered a suitable tracer for atmospheric circulation. It is therefore important to find a model that could relate the observed airborne ^{7}Be activity concentrations, at different latitudes and altitudes, to local meteorological conditions and solar activity variables.

In a preceding study, the weekly surface concentrations of ^{7}Be at six distinct Spanish locations from 2006 to 2021 were analysed using machine learning procedures [10]. The present study builds upon this analysis by extending the scope to observations obtained from three European countries, spanning from the Canary Islands to Finland. This study aims to address the identified shortcomings of the previous work by extending the geographical area of study to a wide range of latitudes and longitudes in order to confirm the validity of the results. The study period, reliant on data availability, covers the years between 2004 and 2023, corresponding to part of the 23rd solar cycle and the entire 24th solar cycle.

2 Data and Methodology

2.1 Data

The data used can be classified into 3 groups according to their origin: observed airborne ^{7}Be activity concentration datasets, surface and upper atmospheric variables and sunspot number observations. The first datasets comprise the variable to be predicted by the random forest (RF) algorithm based model applied at each location, while the latter two act as the features fed to the model.

^{7}Be datasets. The study included a total of 24 measuring sites, dispersed across Spain, France and Finland. The selected network covers a wide range of latitudes and longitudes, as shown in Table 1. These datasets vary in terms of coverage periods and downtimes, they have different availability windows within the 2004-2023 year range.

The high sensibility network of monitoring stations setup by the Spanish Nuclear Safety Council (CSN) analyses, through high-sensitivity measurements, on a weekly basis, the activity concentrations of gamma-emitting isotopes attached to atmospheric aerosols, including ^{7}Be . Six surveillance stations from this Spanish National Radiological Surveillance Network were employed. They are spread along the national territory including the Iberian Peninsula and the Canary Islands. All these activity concentrations since 2000 are publicly available and can be downloaded via a website maintained by the CSN.

The French metrological network for environmental radioactivity centralises all monitoring data for environmental radioactivity in France, approved by the Nuclear Safety and Radiation Protection Authority (ASNR). ASNR manages the radiological monitoring of aerosols at national level in France. Since 2010, all measurements taken in France each year are accessible through a dedicated website. From this database, 10 stations were selected, based on the premise that they should have the longest time span of data collection between 2000 and 2023.

The Radiation and Nuclear Safety Authority of Finland (STUK), monitors radiation in environment in many ways. Its environmental radiological monitoring programme includes the continuous and automatic monitoring of the external dose rate, the monitoring of radioactive substances and total beta activity in outdoor air, as well as the regular radioactivity analysis of radioactive fallout in different type of samples. The activity concentrations of gamma-emitting isotopes attached to atmospheric aerosols from this Finnish monitoring programme can be obtained from an API service hosted by the Finnish Meteorological Institute (FMI). Data from its 8 monitoring stations were employed.

Table 1. Overview of ^{7}Be sampling sites.

Station	Dataset	Period	Longitude	Latitude
CIE-28-PPA-Be7N	Spain	2005-12-26 2021-12-30	-3.726	40.457
ICI-28-PPA-Be7N	Spain	2005-12-26 2021-12-30	-3.725	40.456
UCC-1-PPA-Be7N	Spain	2008-06-24 2021-12-28	-6.349	39.482
ULL-3-PPA-Be7N	Spain	2006-01-01 2021-12-27	-16.291	28.457
UPC-1-PPA-Be7N	Spain	2005-12-28 2021-12-22	2.116	41.384
UPV-1-PPA-Be7N	Spain	2005-12-30 2021-12-28	-2.95	43.261
USE-1-PPA-Be7N	Spain	2005-12-27 2021-12-27	-5.987	37.359
BRT-AS1	France	2010-03-01 2016-08-29	-4.401	48.435
BUG-AS1	France	2010-03-08 2016-08-29	5.29	45.831
CHI-AS1	France	2010-03-01 2016-09-05	0.182	47.206
CIV-AS1	France	2010-03-01 2016-08-29	0.665	46.445
CTN-AS1	France	2010-03-01 2016-08-29	6.245	49.407
GOL-AS1	France	2010-03-01 2016-09-05	0.852	44.115
NIC-AS1	France	2010-03-08 2016-09-05	7.268	43.701
PAL-AS1	France	2010-03-08 2016-09-05	0.629	49.832
SLE-AS1	France	2010-03-08 2016-09-05	1.612	47.717
STA-AS1	France	2010-03-08 2016-09-05	4.775	45.397
Helsinki	Finland	2004-12-18 2022-05-03	25.054	60.209
Imatra	Finland	2005-01-03 2023-11-06	28.827	61.132
Ivalo	Finland	2008-10-23 2023-11-02	27.57	68.64
Kajaani	Finland	2005-01-10 2023-10-23	27.708	64.085
Kotka	Finland	2004-12-27 2023-10-26	26.918	60.483
Kuopio	Finland	2005-02-14 2023-11-13	27.634	62.892
Rovaniemi	Finland	2007-01-02 2023-11-06	25.678	66.513
Sodankylä	Finland	2008-09-08 2023-10-23	26.628	67.365

ERA5. Atmospheric variables were obtained from the ERA5 [5] reanalysis. Produced by the Copernicus Climate Change Service (C3S), and outperforming its predecessor (ERA-Interim), ERA5 is ECMWF's fifth global climate reanalysis. The dataset is constructed through the integration of existing historical observations with modeled estimates of the climate. It utilizes 4D-var data assimilation techniques to process the observations, assimilating them into the CY41R2 cycle of the ECMWF's Integrated Forecast System (IFS) model outputs. ERA5 provides hourly estimates of atmospheric, land, and ocean variables, distributed on a 31 km resolution grid at 137 pressure levels that extend from Earth's surface up to 80 km height.

Table 2 provides an overview of the variables employed in this study. It encompasses both, surface-level and upper-atmosphere variables at a 500 hPa pressure level. These variables were treated as individual time series, with daily frequency centred at midday (12:00), and located at the grid-point closest to each analysed location.

Table 2. Overview of employed ERA5 variables.

Variable	Short name	Unit	Layer
Surface pressure	*sp*	Pa	Surface level
10 m U wind component	*u*10	$\mathrm{m\,s^{-1}}$	Surface level
10 m V wind component	*v*10	$\mathrm{m\,s^{-1}}$	Surface level
2 m dewpoint temperature	*d2m*	K	Surface level
2 m temperature	*t2m*	K	Surface level
Downward UV radiation at surface	*uvb*	$\mathrm{J\,m^{-2}}$	Surface level
Top net solar radiation	*tsr*	$\mathrm{J\,m^{-2}}$	Surface level
Evaporation	*e*	m of water equivalent	Surface level
Total column rain water	*tcrw*	$\mathrm{kg\,m^{-2}}$	Surface level
Total precipitation	*tp*	$\mathrm{kg\,m^{-2}}$	Surface level
Total cloud cover	*tcc*	(-) $[0-1]$	Surface level
Ozone mass mixing ratio	O_3	$\mathrm{kg\,kg^{-1}}$	500 hPa
Vertical velocity	ω	$\mathrm{Pa\,s^{-1}}$	500 hPa

Sunspot Number Dataset. Daily sunspot number observations were utilised as they serve as an indicator of solar activity. This variable is linked to the solar cycle, through which solar radiation, flares, ejection of solar material and sunspot number and size behave synchronised.

This dataset is freely available from the Solar Influences Data Analysis Center (SIDC), hosted by the Royal Observatory of Belgium, and obtained through the Sunspot Index and Long-term Solar Observations (SILSO) network.

2.2 Methodology

Data Format. Each instance of ^{7}Be data used reflects a value accumulated over a period of several days (typically a week) to reduce uncertainties derived from the minimal concentrations intended to be recorded. The recorded value was arbitrarily centred on the middle day of each recording period by the authors, leaving the remaining days blank. Consequently, each ^{7}Be dataset comprises a time series with daily frequency, but mostly empty, containing values only in the middle day of each period. Subsequently, the aforementioned datasets were submitted to a Kolmogorov-Zurbenko (KZ) filter, which separated the high-frequency noise fraction from the main series, while providing daily values to complete each time series.

KZ Filter. The flexibility and robustness of the method render it suitable for a variety of fields, including meteorology and engineering. The technique involves the application of a moving average filter, with a window length of m samples, for p iterations in succession:

$$Y_i = \frac{1}{m} \sum_{j=-\frac{(m-1)}{2}}^{j=\frac{(m-1)}{2}} X_{i+j} \tag{1}$$

The combination of m and p parameters defines the cut-off period of the signal components that the filter allows to pass:

$$mp^{0.5} \leq N \tag{2}$$

This simple formulation embodies a non-parametric low-pass filter capable of dealing with missing data. In this study, the filter was employed with two objectives in mind: first, to separate the main tendencies of ^{7}Be signals from their noise, and second, to complete signals with daily estimates, thereby effectively increasing the resolution from weekly to daily data.

In order to ascertain the optimal m and p configuration, an iterative process was applied to each logarithmic ^{7}Be time series. The combination of every m value ranging from 3 to 45 (with a step of 3 days) and p from 1 to 3 was tested with every time series. The $m = 33$ and $p = 1$ configuration was chosen due to the fact that it exhibited the most appropriate performance in terms of isolating the main fraction from the residual fraction, based on the following conditions:

1. Independence between both fractions, tested by means of correlations and covariances.
2. Normally distributed residual fractions with 0 average. Tested by means of several normality tests, including Pearson Chi Square [4,13], Kolmogorov-Smirnov [9] and Anderson-Darling [11] normality test.

RF Models. Random Forest is a decision tree based supervised machine learning algorithm applicable to classification and regression tasks. The algorithm generates an ensemble of regression trees, where each tree is built using the bagging and feature randomisation techniques. These two techniques introduce diversity and reduce correlation between trees. The output of the model is determined by the average prediction of the ensemble, which reduces the likelihood of over-fitting and improves the generalisation capabilities of the model. A notable aspect of this algorithm is its capacity to provide feature importance scores. The importance of each input feature is estimated based on its contribution to the model's performance. Subsequent to the implementation of the model, its deterioration, in terms of mean decrease in accuracy (MDA), can be computed when input features are shuffled: the mean squared error (MSE) obtained from each tree with the out-of-bag data is compared to the error obtained after permuting each feature. The mean value of these differences over all the trees is obtained, which is then normalised by the standard deviation of the differences. This process enables the user to obtain a table reflecting the importance of each input variable, thereby facilitating understanding of the underlying patterns and relations of data. RF techniques have been extensively described in literature [2,8,12], and have previously been applied to relate ^{7}Be measurements and meteorological variables in Poland [1,14] and Greece [6].

This study focuses on the analysis of the importance tables of each trained model. For the training phase, each dataset was divided into the first 50% for training and the second 50% for validation of the RF models, which was carried out in terms of determination coefficient (R^2) and mean absolute error (MAE) indicators. ^{7}Be time series served as the target dataset, while ERA5 surface and upper datasets, and SSN datasets served as input features. In order to account for the effect of time, lagged signals of ^{7}Be and SSN data were employed. These signals consist of the values corresponding to 7 and 14 days before of each time instance. The models were built using 1000 regression trees and selecting 1/3 of the features at each splitting node, as standard procedures for regression problems indicate. Subsequent to the training of the models, their importance tables were produced and the most relevant 11 variables at each location were identified.

3 Results

3.1 KZ Filter

The $m = 33$ and $p = 1$ configuration works well in general terms. The mean values of the residual fractions range from -0.001 to 0.0035 indicating a null overall contribution to the original time series. The correlation between residues and filtered fraction varies from 0 to 0.06, indicating that both fractions are independent from each other. In terms of normality, each residual fraction is normally or quasi-normally distributed, depending on the presence of punctual outliers in the time series that introduce a small degree of asymmetry. The cut-off period of 33 days makes a clear division of the original time series into 2 fractions independent from each other.

Figure 1 contains two representative cases, of the original time series, the filtered fraction and the residual fraction. Downtimes of each time series can be appreciated.

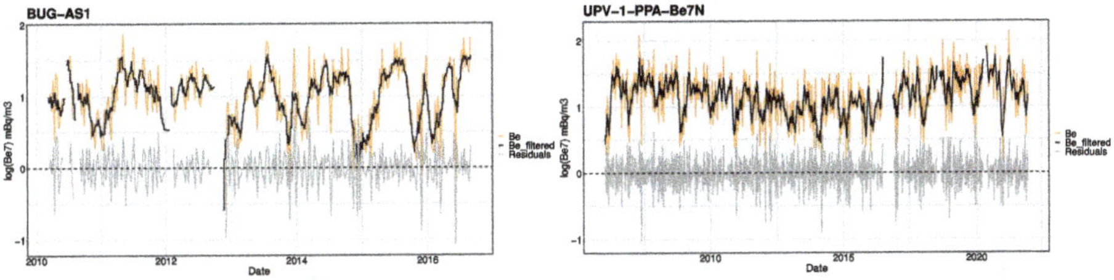

Fig. 1. ^{7}Be concentrations at the BUG-AS1 and UPV-1-PPA-Be7N laboratories(orange). KZ filtered time series are depicted in black. Residuals in grey.(Color figure online)

3.2 Random Forest

As previously stated, 50% of each dataset was utilised for model testing and validation. The outcomes of this split validation technique are displayed in Fig. 2. R^2 represents the proportion of the variability in the target variable (in this case, the ^{7}Be concentration) that is explained by the regression model, with range from 0 to 1. The behaviour of the models trained varies in terms of R^2, especially in the case of French datasets, ranging from 0.55 to 0.91. Those models working with Finnish datasets show better behaviour with R^2 values exceeding 0.75, while those operating on the Spanish datasets exhibit values between 0.84 and 0.9. MAE scores indicate an overall superior performance of the models at Spanish locations (0.058 to 0.093). Conversely, Finnish locations register the highest MAE values (0.127 to 0.135), while the French models lay somewhere in the middle. The models can be grouped by original datasets in terms of performance. Many factors could be responsible for this phenomenon, including the application of different measurement methodologies in origin, variations in downtimes and data availability and The presence of local effects that are not represented by any of the selected input features. At this juncture, the authors lack the requisite information to identify a specific reason to explain these differences. Figure 3 represents the R^2 against geographical location. As can be seen, both longitude and latitude exhibit a lack of statistical significance with regard to model performance.

3.3 Feature Importance

The validity of the feature importance analysis can be accepted for the majority of the models, with a range of R^2 scores between 0.75 and 0.90. However, it is

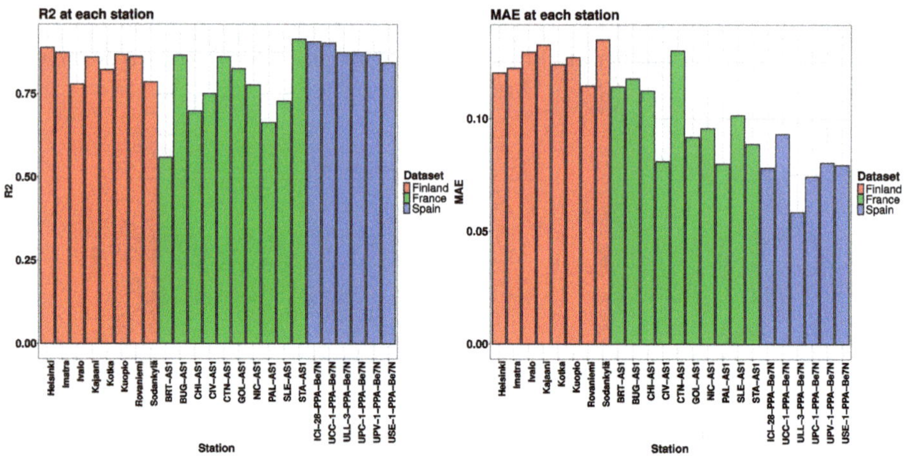

Fig. 2. Validation, R^2 and MAE of each trained model.

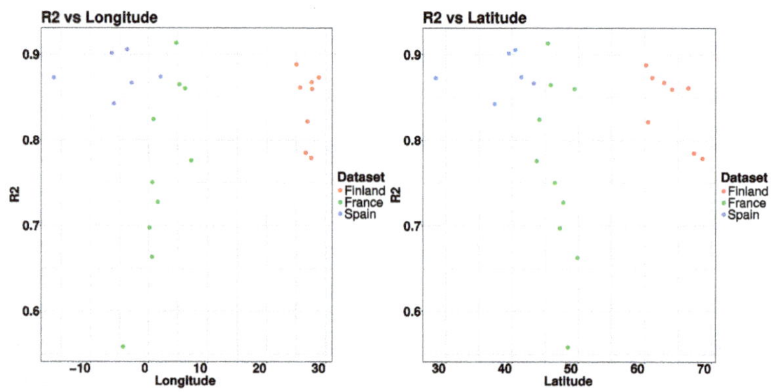

Fig. 3. R^2 of the models against location: longitude and latitude.

important to note that this study encompasses the results for all the trained models, including those with poor performance. While these models may not be suitable for extracting ^{7}Be concentration behaviour in the atmosphere, they could be useful for improving the applied methodology.

Figure 4 summarises the results of the feature importance analysis. Only the 6 most relevant features at each location are displayed. The classification was made in terms of MDA, meaning that those features with a higher score induce higher errors on the model when being permuted. Consequently, these features are deemed more pertinent to the model's performance. As can be observed the most relevant variable in every case is the 1 week delayed ^{7}Be signal with MDA values over 70%, followed by the 2 week delayed signal with up to 47%. This means that models are predicting the ^{7}Be signal based on their own past values, which can be related to the memory of the system or the autocorrelation of time series. The remaining most relevant features can be classified in two groups:

- Solar activity related variables: daily sunspot number (20% to 47% MDA) and its lagged signals (10% to 50% MDA) play a relevant role the performance of most models. This variable can be used as a proxy of solar activity, which is highly related to the ^{7}Be generation mechanism. Total solar radiation and downward UV radiation at the surface may also classified in this group, though they are dependent on other atmospheric phenomena such as total cloud cover, which makes them unsuitable as direct proxies of solar activity.
- Precipitation related variables: evaporation (27% to 47% MDA), surface pressure (33% to 41% MDA), 2 m temperature (22% to 45% MDA) and 2 m dew-point temperature (28% to 43% MDA) are all variables related to precipitation mechanisms of water, and linked to them aerosols and airborne particulate matter.

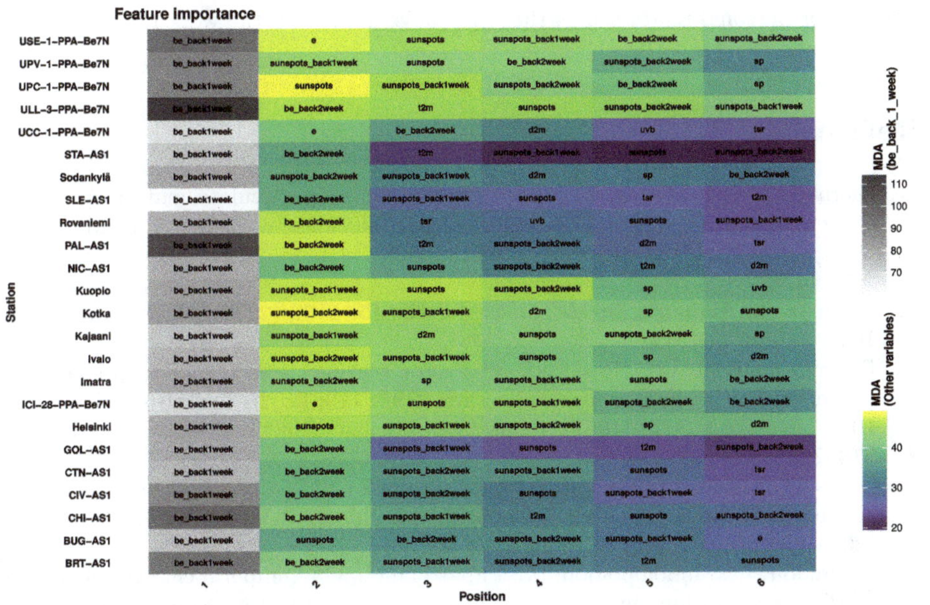

Fig. 4. Feature importance of each model, top 6 more relevant features. In order to enhance the interpretability of the graph, a dual-scale approach has been employed, with one scale exclusively associated with the *be_back1week* variable and the other associated with the remaining variables.

4 Conclusion

This study analysed the ^{7}Be radionuclide time series, measured at 24 measuring laboratories dispersed across Spain, France and Finland, corresponding to the

2004–2023 period together with surface and upper atmospheric variables and sunspot number observations. Applying the combined approach of the KZ filter followed by the application of RF models, to all time series, independent analysis of both the seasonal component and the noisiest fraction of the observations was enabled. This analysis concludes that models can predict the ^{7}Be signal based firstly on their own past values, which can be related to the memory of the system or the autocorrelation of time series, secondly on solar activity related variables and thirdly on rainfall related variables. In the future, similar machine learning analysis combined with cluster analysis will be used for non-cosmogenic, terrestrial, radionuclides, such as ^{210}Pb and ^{40}K in aerosols.

Acknowledgments. This study was funded by TED2021-132109B- C21 (HOBE) MCIN/AEI/10.13039/501100011033 and is part of project PID2020-116153RB-I00 The authors also acknowledge funding from SENyRAD (IT1694-22), a research group at the Basque University System of the Basque Government.

Disclosure of Interests. The authors declare that they have no competing financial interests or personal relationships that may have influenced the work reported in this study.

References

1. Biernacik, D., Zalewska, T.: ^{7}Be, ^{210}pb, Airborne particulate matter and PM10 concentrations in relation to meteorological conditions in southern Poland in 1998–2016. J. Environ. Radioact. **259**, 107122 (2023). https://doi.org/10.1016/j.jenvrad.2023.107122
2. Breiman, L.: Random forests. Mach. Learn. **45**, 5–32 (2001). https://doi.org/10.1023/A:1010933404324
3. Długosz-Lisiecka, M., Nowański, S.: Factor regression as a tool to predict Be-7 concentration in the air. Atmos. Environ. **297**, 119612 (2023). https://doi.org/10.1016/j.atmosenv.2023.119612, https://www.sciencedirect.com/science/article/pii/S1352231023000389
4. Gross, J., Ligges, U.: Nortest: tests for normality (2015). R package version 1.0-4
5. Hersbach, H., et al.: The Era5 global reanalysis. Q. J. Royal Meteorol. Soc. **146** (2020). https://doi.org/10.1002/qj.3803
6. Ioannidou, A., Manolopoulou, M., Papastefanou, C.: Temporal changes of ^{7}Be and ^{210}Pb concentrations in surface air at temperate latitudes (40 N). Appl. Radiat. Isot. **63**(2), 277–284 (2005). https://doi.org/10.1016/j.apradiso.2005.03.010
7. Kong, Y., Lee, O.S., Yung, C.: Study of the naturally occurring radionuclide beryllium-7 (Be-7) in Hong Kong. J. Environ. Radioact. **246**, 106850 (2022). https://doi.org/10.1016/j.jenvrad.2022.106850. https://www.sciencedirect.com/science/article/pii/S0265931X22000406
8. Liaw, A., Wiener, M., et al.: Classification and regression by randomForest. R news **2**(3), 18–22 (2002)
9. Massey, F.J.: The Kolmogorov-Smirnov test for goodness of fit. J. Am. Stat. Assoc. **46**(253), 68–78 (1951). http://www.jstor.org/stable/2280095
10. Nafarrate, A., et al.: Applying the kolmogorov–zurbenko filter followed by random forest models to 7Be observations in Spain (2006–2021). Heliyon **10**(9) (2024)

11. Scholz, F.W., Stephens, M.A.: K-sample Anderson-darling tests. J. Am. Stat. Assoc. **82**(399), 918–924 (1987). http://www.jstor.org/stable/2288805
12. Siroky, D., et al.: Navigating random forests and related advances in algorithmic modeling. Stat. Surv. **3**, 147–163 (2009). https://doi.org/10.1214/07-SS033
13. Thode, H.: Testing for normality marcel Dekker. Inc. New York, pp. 99–123 (2002). https://doi.org/10.1201/9780203910894
14. Zalewska, T., Biernacik, D., Marosz, M.: Correlations between ^{7}Be, ^{210}Pb, dust and PM10 concentrations in relation to meteorological conditions in northern Poland in 1998–2018. J. Environ. Radioact. **228**, 106526 (2021). https://doi.org/10.1016/j.jenvrad.2020.106526

Classification of Multimodal Biomedical Data

Dariusz Jankowski(✉) and Jakub Kardela

Department of Systems and Computer Networks, Faculty of Information and Communication Technology, Wrocław University of Science and Technology, Wybrzeże Wyspiańskiego 27, 50-370 Wrocław, Poland
dariusz.jankowski@pwr.edu.pl

Abstract. This paper explores deep learning models for classifying multimodal biomedical data using various fusion strategies. By combining medical images and metadata, the study evaluates the impact of different fusion approaches on classification performance. The findings highlight the potential of advanced fusion methods to enhance diagnostic accuracy in medical applications.

Keywords: Multimodal Data Fusion · Biomedical Classification · Autoencoder-Based Feature Extraction · Deep Neural Networks

1 Introduction

This study aims to design and implement deep neural network models utilizing selected fusion strategies and compare their effectiveness in classifying multimodal biomedical data. The efficiency of these fusion methods, measured by classification quality, will be evaluated using specific metrics relevant to medical problems. Research questions will address the impact of applied approaches on classification quality measures. The study will employ an experimental protocol, and the results will be subjected to statistical analysis to determine the significance of differences between the methods.

The motivation for this research stems from the growing potential of multimodal data classification, evidenced by the increasing number of multimodal datasets and applications. In medicine, multimodal learning can enhance disease diagnosis and treatment selection. The subsequent chapters will explore related topics, presenting a literature review on multimodal data classification, taxonomy, and the primary division criteria of fusion strategies. The study will also demonstrate the data set, experimental plan, deep neural network models implementing selected fusion strategies, and the experimental environment. Finally, the quality of the solutions will be assessed using statistical tests, and the conclusions will suggest future research directions.

The following are the key contributions of this paper:

1. Development of deep neural network models implementing various multimodal fusion strategies.

P. Trajdos and R. Burduk (Eds.): CORES 2025, LNNS 1557, pp. 54–65, 2025.
https://doi.org/10.1007/978-3-032-01773-4_6

2. Comprehensive comparison of the classification quality of these models using specific medical metrics.
3. Detailed analysis of the impact of different fusion strategies on classification performance.
4. Application of a robust experimental protocol, including statistical validation, to ensure the reliability of results.
5. Insights into the practical applications and potential improvements in multimodal data classification for medical use.

The rest of this paper is organized as follows: Sect. 2 reviews related work on multimodal fusion of biomedical data. Section 3 describes the dataset, methods, and experimental results. Section 4 concludes with key findings and future directions.

2 Related Work

The integration of heterogeneous biomedical data modalities, such as dermatological images, clinical scans, and patient metadata, offers significant potential for improving diagnostic accuracy and clinical decision support. However, the effectiveness of these models heavily depends on the fusion strategy used to combine information from multiple data sources. This review analyzes comparative studies and recent research across five categories of fusion: early fusion, intermediate fusion (extreme and joint), and late fusion.

2.1 Early Fusion

Early fusion, or feature-level fusion, concatenates raw or low-level features from each modality before passing them into a joint model. This approach is widely used due to its simplicity and ability to model cross-modal interactions early in the pipeline.

- *Strengths*: Captures fine-grained feature dependencies; efficient during inference.
- *Limitations*: Prone to overfitting, especially with heterogeneous data dimensionality or small datasets.

In dermatological classification tasks, [2] employed early fusion by combining dermoscopic image embeddings with patient metadata. While results showed improvement over unimodal baselines, the approach was sensitive to missing data. Similarly, [3] applied early fusion in skin cancer classification, noting decreased robustness when clinical text was noisy or incomplete.

2.2 Intermediate Fusion

Intermediate fusion methods process each modality separately before fusing at a deeper representation level. This allows independent learning of modality-specific representations while still modeling their joint semantics. Two subtypes are increasingly distinguished:

Extreme Intermediate Fusion. This strategy fuses mid-level features at later network layers, typically using multi-branch architectures with fusion at penultimate stages. In [4], transformers were used with modality-specific encoders for clinical and dermoscopic images, merging at deeper layers. This architecture achieved state-of-the-art accuracy on the ISIC dataset.

- *Strengths*: Balances modularity with integration.
- *Limitations*: Sensitive to synchronization errors between modalities.

Joint Intermediate Fusion. This method involves learning a shared representation through alignment or co-attention mechanisms. For instance, [5] introduced a co-attention mechanism to align dermatological images with structured metadata, enhancing lesion classification. The joint embedding space facilitated robustness to modality imbalance.

Tensor fusion networks (TFN) and modality-invariant representations are also part of this class. These methods provide stronger modeling capacity at the cost of increased complexity.

2.3 Late Fusion

Late fusion, or decision-level fusion, aggregates outputs from independent modality-specific classifiers. Popular in ensemble models and scenarios with missing or asynchronous data, this method is modular and interpretable.

- *Strengths*: High modularity; easier to handle missing modalities.
- *Limitations*: Fails to capture deep cross-modal feature interactions.

In [6] authors utilized separate CNNs for dermoscopic images and patient information, with outputs combined using weighted averaging. Although this method showed competitive performance, its gains plateaued compared to intermediate fusion techniques. A similar conclusion was reached in [7], where late fusion failed to exploit complementary information across modalities.

2.4 Comparative Analyses and Controlled Benchmarks

Few studies provide standardized comparisons across fusion types under identical experimental conditions—a notable gap. However, some exceptions exist:

Article [8] compared early, joint intermediate, and late fusion in a controlled setting using a multimodal dermatological dataset. Hybrid (joint intermediate) fusion yielded the highest F1-score (0.89), outperforming early (0.82) and late fusion (0.84).

Authors in [9] introduced a benchmarking protocol for medical multimodal fusion, suggesting that joint intermediate fusion consistently achieves better generalization, especially under distribution shift and missing modality conditions.

Statistical robustness was emphasized in [10], where bootstrapped confidence intervals showed significantly narrower variance in joint fusion models.

These findings align with the hypothesis that hybrid fusion approaches, especially joint intermediate methods leveraging attention or latent alignment, outperform others in terms of both accuracy and robustness.

2.5 Interpretability and Clinical Usability

Interpretability remains a concern, particularly for early and joint fusion methods. While late fusion allows clear attribution of performance to individual modalities, intermediate fusion requires more advanced explainability techniques (e.g., SHAP or attention visualization) [11].

Ethical considerations in multimodal biomedical AI also arise due to potential biases in metadata (e.g., ethnicity, age), necessitating fairness-aware fusion techniques, as explored in [12].

3 Experiments

3.1 Dataset Description

The utilized dataset contains information on 1,011 skin lesions described through three modalities: dermatological images, clinical images, and metadata [1]. Metadata, which consist of categorical variables, includes the 7-point scale evaluation, type of pigment network, streak shape, pigmentation shape, regression structure type, shape of dots and globules, presence of blue-white veil, vascular structure shape, diagnostic difficulty level, elevation shape, lesion location, and patient gender. All images resolution is 768 by 512 pixels and are described through three color channels. This dataset comprises multiple classes but can be reduced to two classes: benign and malignant lesions, the latter also referred to as melanoma. There are 252 positive cases and 759 negative cases, indicating an imbalance in the dataset.

The experiments were conducted in the Google Colaboratory environment, which allows running code written in Python. A computational GPU unit with 40 GB of VRAM and 83.5 GB of system RAM was utilized.

3.2 Experiments

The aim of the experiments is to investigate the impact of applied multimodal data fusion methods on classification quality using deep neural networks. The following models implementing selected feature-level fusion methods will be compared: 1) model using **direct modeling**, which takes concatenated features from all modalities as input, 2) similar model that uses concatenated features extracted through a **multimodal autoencoder**, and 3) model with **intermediate late fusion**, where feature concatenation occurs within the model architecture. Classification quality will be measured using metrics such as balanced accuracy, precision, and recall.

Defining the research goal and problem area enabled the formulation of the following research questions:

1. What is the impact of applying class weighting on classification quality in the case of an imbalanced multimodal dataset?
2. What is the impact of the bottleneck size in the multimodal autoencoder, used to obtain a shared representation, on classification quality?
3. Which fusion method results in the highest classification quality?

The experimental protocol employs a stratified 5-fold cross-validation repeated twice. To verify the hypotheses regarding whether the mean values of a given metric differ statistically based on the applied method, paired Student's t-tests were performed.

The data were preprocessed as follows: images were rescaled to a size of 192 by 128 pixels, and pixel values were normalized to a range of 0 to 1. The resolution reduction was dictated by hardware limitations. Additionally, metadata features were encoded using one-hot encoding, increasing the number of metadata features to 59.

Direct Modeling. Features from medical images were extracted using the VGG16 model. For further dimensionality reduction, the extracted features were subjected to principal component analysis (PCA). The number of retained components was determined so that the cumulative explained variance would amount to 80%. This reduced the number of features of dermatological images to approximately 114 and features of clinical images to approximately 105. Metadata was also subjected to PCA, ensuring that the retained components explained 95% of the variance, which reduced their dimensionality to 29 features.

The classifier model was designed as a simple artificial neural network with a single hidden dense layer. This architecture has three inputs, one for each modality: features of dermatological images, features of clinical images, and metadata features. These are then combined by a concatenation layer. Following this is a dropout layer with a dropout rate of 0.5. Next, a dense layer with 32 neurons is used to reduce dimensionality. This layer employs the ReLU activation function. The final layer is an output layer with a single neuron, which returns the probability of class membership. A sigmoid activation function is used here to normalize the output value to the range of 0 to 1. This approach schema is illustrated in Fig. 1 and the architecture of this network is shown in Fig. 2. During model compilation, binary cross-entropy was used as the loss function, and the Adam optimizer was applied with a learning rate of 0.0001.

Multimodal Autoencoder. In this approach, features were obtained as a shared latent representation of all modalities. This representation was determined using a multimodal autoencoder, whose architecture has three inputs: dermatological images, clinical images, and metadata. The shared features obtained were then used as input for a separate classifier model with a structure analogous to the one used in the previous approach. A key element of the autoencoder's

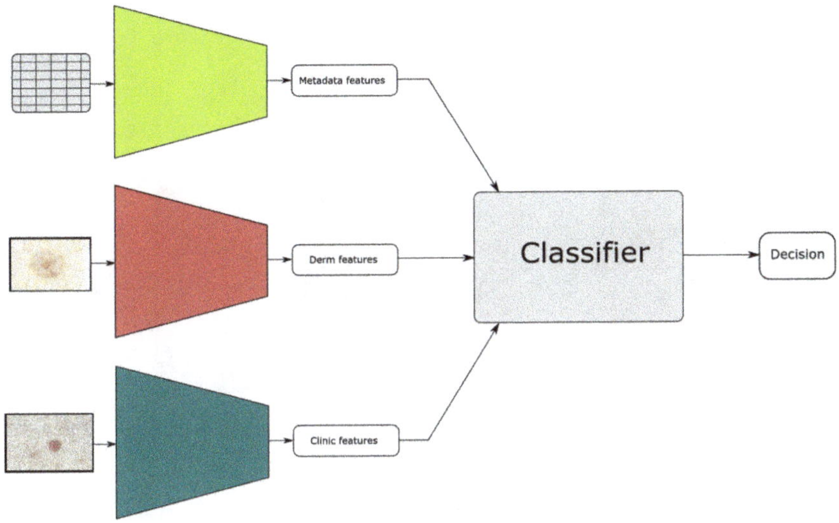

Fig. 1. Classifier Diagram

architecture is the bottleneck size, which determines the number of features to which the information is compressed. A larger bottleneck size allows for better reconstruction of the input at the output, while an excessively large size may lead to overfitting during the training of the classifier model. In the study, classification quality was compared based on features extracted from autoencoders with different bottleneck sizes—64, 256, and 512. The optimal variant was selected for the final comparison with other methods.

The autoencoder architecture is divided into two main parts: an encoder and a decoder. A single dense layer between them serves as the bottleneck, concentrating compressed information. The encoder and decoder are mirror images in structure, differing only in the direction of information flow and the fact that the decoder performs operations inverse to the encoder. The autoencoder has three inputs (and corresponding outputs): dermatological images, clinical images, and metadata. The autoencoder schema is illustrated in Fig. 3. The sections corresponding to each input are responsible for dimensionality reduction and generating increasingly refined edge representations. For image data, these sections consist of four consecutive segments of convolutional layers with a kernel size of 3×3 and max pooling layers with dimensions of 2×2. The convolutional layers have 16, 32, 64, and 64 neurons, respectively, with ReLU activation functions applied in each layer. Following this is a flattened layer, which converts feature maps into a flat vector. Next is a dense layer with a ReLU activation function that reduces dimensionality to 96 neurons. Subsequently, a concatenation layer combines feature vectors of dermatological images, clinical images, and metadata. The resulting vector serves as the input to the bottleneck layer, a dense layer with a ReLU activation function. The number of neurons in this

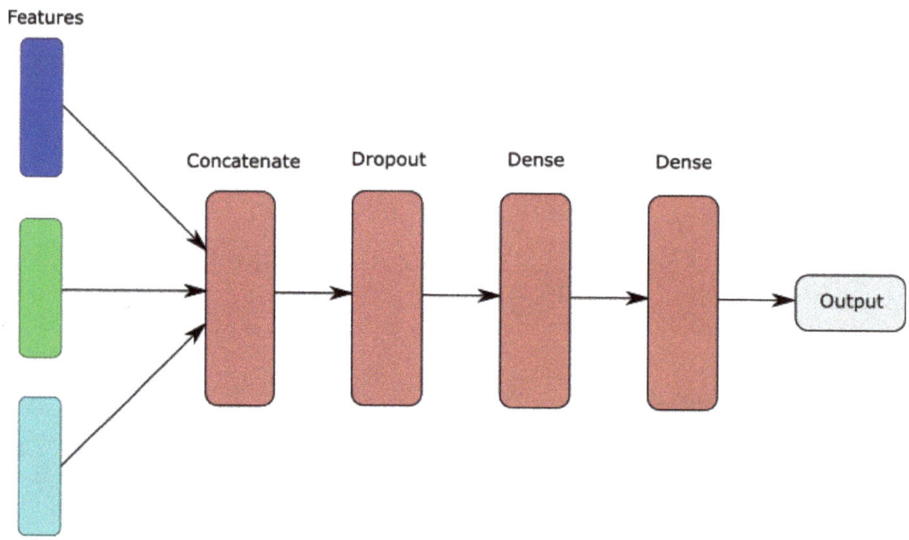

Fig. 2. Classifier Architecture Diagram

layer depends on the autoencoder's bottleneck parameter. The remaining parts of the architecture belong to the decoder. Structurally, these layers are analogous to those in the encoder but are arranged in reverse order. The operations performed by the decoder are also the opposite. Instead of the flattened layer, a reshaped layer transforms the flat vector back into feature maps. Alternating layers include convolutional layers and up-sampling layers (2×2). The final convolutional layer uses a sigmoid activation function to reconstruct images, with pixel values normalized between 0 and 1. The sigmoid activation function is also applied in the dense layer responsible for reconstructing metadata outputs. During model compilation, the Adam optimizer was used, and the mean squared error was applied as the loss function for each output. This metric evaluates how well the autoencoder reconstructs the input data it receives.

The features obtained in this manner, as in the previous approach, were used as input for the final classifier. The classifier was compiled using the same hyperparameters as in the direct modelling strategy.

Intermediate Late Fusion. In this strategy, a neural network model was built that acts both as a feature extractor for individual modalities and as a classifier. The model has three inputs, one for each modality: dermatological images, clinical images, and metadata. The architecture of this network is analogous to the encoder used in the autoencoder method. An additional component consists of classifier layers, which were also used for decision-making in the two previous approaches. Furthermore, dropout layers with a dropout rate of 0.5 were added after the third and fourth segments of the convolutional and pooling layers to counteract overfitting. The architecture of this model is illustrated in Fig. 4. The

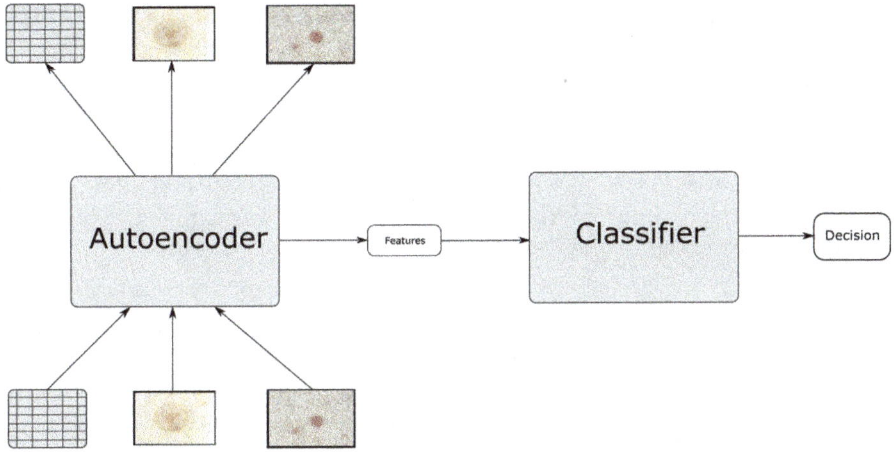

Fig. 3. Autoencoder Diagram

Adam optimizer with a learning rate of 0.0001 was used during model compilation, and binary cross-entropy was adopted as the loss function.

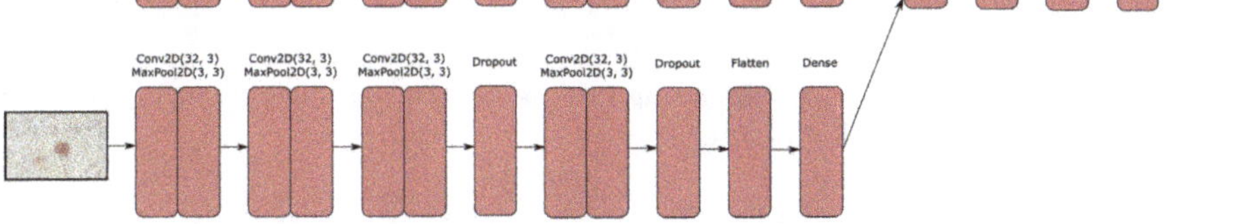

Fig. 4. Architecture Diagram of Intermediate Late Fusion

Due to the imbalanced dataset, class weighting was applied in cases when the task involved training networks for classification. Given the imbalance ratio of 1:3, the minority class was assigned a weight of 3, while the majority class was assigned a weight of 1. This approach ensures that the model places greater emphasis on correctly classifying the minority class, which is critical in the task of identifying affected individuals. Additionally, early stopping was employed during training to halt the process when overfitting occurred. The training was terminated if the validation loss did not improve for 25 epochs, and the model state corresponding to the minimum validation loss was retained. For all models, the batch size was set to 4.

3.3 Results and Discussion

In an **Experiment 1**, the classification quality of the direct modelling method was compared depending on whether class weighting was applied. The results obtained are presented in the Table 1.

Table 1. The classification quality of direct modelling depending on the application of class weighting (the best results have been highlighted).

Class Weighting	Balanced Accuracy	Precision	Recall
No	0,770	**0,803**	0,588
Yes	0,823	0,661	**0,780**

Precision decreased, although recall increased. However, the improvement in recall outweighs the decrease in precision, which is also reflected in the overall rise in balanced accuracy. *Paired Student's t-tests* indicate a statistically significant difference for both precision and recall depending on the use of class weighting. While balanced accuracy also increased, this difference was not statistically significant. The metrics that improved are crucial for the task of identifying affected individuals; hence, class weighting was applied in experiments 2 and 3.

In the subsequent **Experiment 2**, aimed at comparing the classification quality of the autoencoder method depending on the bottleneck size, the results obtained are presented in the Table 2.

Table 2. Classification quality depending on the bottleneck size of the autoencoder (the best results have been highlighted).

Bottleneck Size	Balanced Accuracy	Precision	Recall
64	0,843	**0,737**	0,780
256	0,862	0,715	0,836
512	**0,883**	0,733	**0,872**

A clear relationship is observed between the metric values and the bottleneck sizes of the autoencoder. In general, the values of all metrics increase as the bottleneck size grows. An exception occurs with a bottleneck size of 64, where precision is higher than a bottleneck size of 256. This may be because class weighting was not applied during the training of the autoencoders, leading the autoencoder with a small bottleneck size to inadequately encode minority class objects. Apart from this exception, the increase in metric values with larger bottleneck sizes could be attributed to the greater amount of information that can be encoded in a larger bottleneck. In the matter of recall, the difference between the values for bottleneck sizes of 512 and 256 is smaller than the difference between sizes 256 and 64. This may indicate that a certain bottleneck size is sufficient to encode enough information for the classifier model to effectively learn the problem.

Paired Student's t-tests indicated that, in terms of balanced accuracy, the approach using an autoencoder with a bottleneck size of 512 neurons was statistically the best for the classification task. However, in the matter of precision, there were no statistically significant differences among the compared models. Similarly, referring to recall, a statistically significant difference was observed only between the bottleneck sizes of 512 neurons and 64 neurons.

Experiment 3 focused on further comparisons, based on the optimal bottleneck size of 512 neurons, which was selected after analyzing the results of the study on the impact of the autoencoder bottleneck size on classification quality. The results of this approach, along with the results of other multimodal data fusion methods, are summarized in Table 3. To ensure a clearer presentation of the research results and statistical tests, abbreviations were used for the compared methods: direct modelling - DM, the autoencoder method - AE, and intermediate late fusion - ILF.

Table 3. Classification quality depending on the fusion method (the best results have been highlighted).

Fusion Method	Balanced Accuracy	Precision	Recall
MB	0,823	0,661	0,780
AE	**0,883**	**0,733**	**0,872**
KFP	0,863	0,721	0,836

The table shows that the metric values for the autoencoder and intermediate late fusion approaches are higher than the corresponding measures for direct modelling. The results of both methods are very similar, but the autoencoder approach demonstrates slightly higher metrics. For better visualization of the performance of the compared models, their averaged Receiver Operating Characteristic (ROC) curves are presented in Fig. 5.

In this figure, it is visually apparent that the curves of the autoencoder approach and intermediate late fusion are closest to the top-left corner. In contrast, the curve of direct modelling is significantly farther away, indicating lower quality for this approach.

Paired Student's t-tests indicated that, in terms of balanced accuracy, both the autoencoder method and intermediate late fusion were statistically significantly better for the classification task than direct modeling. Similarly, when precision was the criterion, both models were statistically significantly better

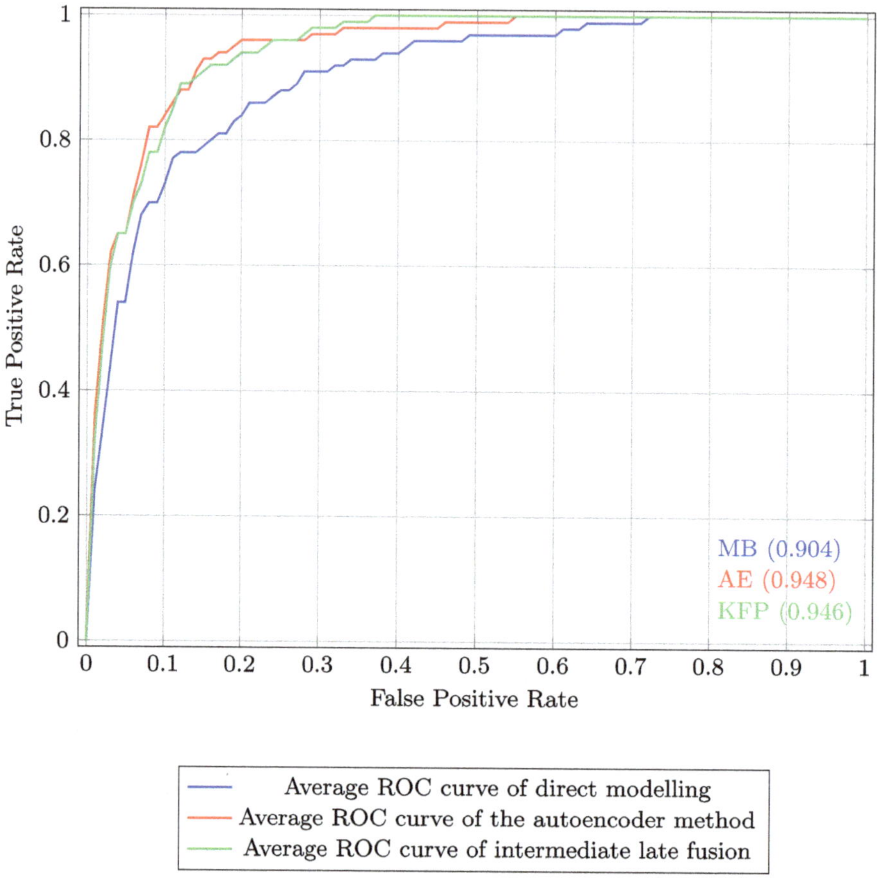

Fig. 5. Average ROC curves of the compared fusion methods.

than direct modelling. In the matter of recall, only the autoencoder approach was statistically significantly better than direct modelling. However, none of the measured metrics did the autoencoder method or intermediate late fusion show a statistically significant advantage.

4 Conclusion and Future Works

This study explored the performance of various multimodal data fusion methods in the task of biomedical classification using deep neural networks. The approaches compared included direct modelling, the use of a multimodal autoencoder, and intermediate late fusion. The results demonstrate that both the autoencoder and intermediate late fusion methods significantly outperform direct modelling regarding balanced accuracy, precision, and recall. While the differences between the autoencoder and intermediate late fusion methods were not statistically significant, the autoencoder approach achieved slightly higher metrics across all categories.

The findings highlight the potential of leveraging advanced feature extraction and fusion techniques to enhance the classification of multimodal biomedical datasets. Particularly, the autoencoder method showed strong performance in handling imbalanced datasets and encoding comprehensive shared representations, making it a promising candidate for future applications in medical diagnostics.

References

1. Kawahara, J., Daneshvar, S., Argenziano, G., Hamarneh, G.: Seven-point checklist and skin lesion classification using multitask multimodal neural nets. IEEE J. Biomed. Health Inform. **23**(2), 538–546 (2019). https://doi.org/10.1109/JBHI.2018.2824327
2. Esteva, A., Kuprel, B., Novoa, R.A., et al.: Dermatologist-level classification of skin cancer with deep neural networks. Nature **542**(7639), 115–118 (2017)
3. Ghosh, R., Liu, A., Yoon, J.: Early fusion improves multimodal skin cancer diagnosis. In: Proceedings of the MICCAI (2021)
4. Valle, E., Silva, F., Santana, J.: SkinTransformer: a hierarchical vision transformer for skin lesion classification. Comput. Med. Imaging Graph. **92**, 101967 (2021)
5. Huang, Z., Chen, W., Zhang, X.: Joint image and metadata fusion for skin disease classification. IEEE Trans. Med. Imaging **40**(10), 2758–2770 (2021)
6. Serte, S., Demirel, H.: A deep learning approach for skin lesion classification with metadata fusion. In: International Conference of Medical Imaging, pp. 1–5. IEEE (2019)
7. Rajpurkar, P., Irvin, J., Ball, R., et al.: CheXMix: multimodal learning for medical diagnosis with chest X-ray and structured data. Nat. Commun. **13**, 1–13 (2022)
8. Ma, J., Zhou, Y., Sun, Q.: Multimodal fusion of dermoscopic images and clinical metadata for skin lesion classification. Pattern Recognit. **114**, 107828 (2021)
9. Zhao, L., Li, J., Song, J.: A taxonomy and evaluation framework for multimodal fusion in biomedical applications. IEEE J. Biomed. Health Inform. **26**(11), 5438–5451 (2022)
10. Yu, T., Wang, Z., Liu, W.: Robust fusion strategies for multimodal medical diagnosis: statistical perspectives. Med. Image Anal. **85**, 102749 (2023)
11. Holzinger, A., Langs, G., Denk, H., et al.: Towards robust, explainable and trustworthy AI for medical imaging. Nat. Rev. Cancer **22**(1), 35–50 (2022)
12. Seyyed-Kalantari, L., Zhang, I.Y., McDermott, M., et al.: CheXclusion: fairness gaps in deep chest X-ray classifiers. In: Proceedings of the ACM Conference Fairness, Accountability, and Transparency (FAccT). ACM (2020)

Graph-Based Learning for Electrical Quantities Prediction in Energy Communities

Lucia Porlan-Ferrando, J. David Nuñez-Gonzalez(✉), and Iker Aretxabaleta

Engineering School of Gipuzkoa - Eibar, University of Basque Country UPV/EHU, Leioa, Spain

josedavid.nunez@ehu.eus

Abstract. This study examines the design and management of energy communities in rural environments, where the geographic dispersion of households and the limitations of conventional electrical infrastructure make local networks a more viable and sustainable solution. Energy communities enable residential and commercial users to take on a dual role as consumers and prosumers, maximizing the integration of renewable sources such as solar and wind energy. However, the inherent variability of these sources, combined with demand patterns, introduces significant technical challenges in terms of system stability and efficiency. The proposed methodology is structured in two main phases. In the first phase, a rural energy community will be modeled using Simulink, taking into account the consumption and generation profiles of each household, as well as their geographic distribution, to incorporate conductor loss factors and regulatory constraints on voltage, current, and power. This model will generate databases representing energy generation and consumption over a daily cycle. The resulting analysis will consider current regulations and seek to optimize energy self-sufficiency by minimizing dependence on the conventional power grid and maximizing the utilization of surplus energy from interconnected households. The study integrates simulations with machine learning techniques to predict grid intensity one hour in advance, aiming to mitigate overcurrent occurrences and enhance the system's operational stability.

1 Introduction

In recent decades, the transition towards a more sustainable energy system has become a global priority, driven by the need to reduce dependence on fossil fuels and mitigate the effects of climate change. Within this context, energy communities have emerged as an innovative and efficient solution for integrating renewable energy sources into decentralized electrical systems. These communities enable various stakeholders, such as households and small businesses, to interact as energy consumers and/or producers, thereby maximizing the utilization of local renewable resources such as solar or wind energy [7].

P. Trajdos and R. Burduk (Eds.): CORES 2025, LNNS 1557, pp. 66–74, 2025.
https://doi.org/10.1007/978-3-032-01773-4_7

The concept of an energy community is based on collaboration and the optimal utilization of local energy generation and consumption, promoting energy independence, cost reduction, and minimization of transmission losses [9]. However, the design and management of these communities pose significant technical challenges. The inherent variability of renewable sources, such as solar and wind generation, combined with the heterogeneous consumption patterns of households, introduces considerable uncertainties in the energy balance of the grid [3]. Additionally, factors such as conductor losses, regulatory limits on voltage, current, and power, and the spatial distribution of households must be considered to ensure the efficient and stable operation of these communities [9].

In rural areas, where households tend to be more dispersed and conventional electrical infrastructure faces greater limitations, energy communities become even more relevant. In these regions, extending cabling from the main electrical grid to each household can be technically and economically unfeasible. Energy communities offer a sustainable and practical alternative by enabling the creation of local networks in which users share energy resources, maximizing distributed generation while minimizing the need for costly infrastructure. Furthermore, this approach can ensure a more stable and autonomous electricity supply [7].

This study proposes a two-stage approach to address these challenges. The first stage involves the detailed modeling of a rural energy community using Simulink. This model will represent an electrical network in which each household has a specific consumption profile and, if acting as a generator, a production profile based on renewable technologies. Key considerations will also be included, such as the distance between households, which affects conductor losses, and regulatory limits. Throughout the simulation, data will be collected to generate databases that reflect the dynamics of energy consumption and generation within the community. [2].

The second phase of the study will focus on predictive analysis using graph-based machine learning methods. This approach will enable the identification and management of overcurrent occurrences within the energy community. The model will predict current intensity one hour in advance to ensure compliance with regulatory limits and mitigate potential damage to the grid. This strategy not only enhances system stability but also promotes greater energy self-sufficiency within the community.

In summary, this study integrates advanced simulation tools and machine learning techniques to model, analyze, and optimize rural energy communities, contributing to the development of scalable and sustainable solutions for energy systems.

2 State of the Art

The concept of energy communities has evolved significantly in recent decades as a viable solution for the transition toward a decentralized and sustainable energy model. These communities have been the focus of numerous studies addressing their implementation, management, and optimization, employing various approaches, including energy modeling, the analysis of renewable generation variability, and the use of machine learning algorithms for system prediction and optimization.

The simulation and modeling of energy communities have been extensively studied in the literature, utilizing tools such as Simulink, HOMER, and OpenDSS. Various studies have demonstrated the importance of modeling both energy generation and consumption with high temporal and spatial resolution. For instance, Luthander et al. (2015) analyzed photovoltaic self-management in buildings, highlighting the need for strategies to maximize self-consumption efficiency [5]. The integration of renewable energy sources in energy communities has also been addressed in studies emphasizing the necessity of considering the intermittency of photovoltaic and wind generation. In this regard, Etxeberria et al. (2012) proposed hybrid storage systems to enhance the efficient integration of renewable energies into microgrids [4].

In rural areas, where conventional electrical infrastructure is limited, implementing energy communities presents a unique challenge. Recent studies have identified technical and economic barriers, such as high energy transmission losses and the difficulty of synchronizing electricity supply and demand in isolated systems (Sevgen et al., 2025). To mitigate these challenges, research has explored strategies to improve energy autonomy [6]. Moreover, artificial intelligence-based optimization strategies have proven effective in demand management and planning, maximizing the utilization of local resources (Bode et al., 2015) [1].

The application of machine learning (ML) algorithms in the management of energy communities has gained significant relevance in recent years. Specifically, approaches based on neural networks, clustering, and graph models have been employed for consumption and generation prediction as well as for optimizing the topology of the electrical network. Jiajun et al. (2021) implemented reinforcement learning strategies to optimize energy storage control in buildings [1], while Zhou et al. (2022) utilized graph-based neural networks to enhance energy management within communities [10]. Recent research has demonstrated that graph-based learning models can improve the prediction of energy generation and consumption in energy communities, enabling a more efficient resource management approach and promoting sustainability. This methodology considers the complex interactions among different stakeholders within the community, improving the accuracy of estimations and facilitating decision-making in energy distribution (Porlan et al., 2024) [8].

The state of the art in energy communities reflects a growing interest in the application of advanced technologies to enhance the efficiency and sustainability of these systems. The combination of detailed simulations with machine learning models has proven to be a promising strategy for optimizing energy management and facilitating the integration of new households into rural communities. However, challenges related to scalability, regulation, and the economic feasibility of these solutions remain, highlighting the need for further research in these areas.

3 Proposed Experiment

As previously mentioned, this article is structured into two sections: one focused on simulation and the other on prediction using graph-based Machine Learning models. The following sections provide a detailed description of both stages.

3.1 Simulation

A computational model has been developed in Simulink to simulate the energy community and analyze its electrical behavior. The microgrid consists of:

- 52 households with electrical consumption and renewable generation (photovoltaic or wind).
- 6 substations responsible for energy distribution.
- Main electrical grid, which interconnects all elements and ensures system stability.

The simulation uses daily consumption and production curves as input, representing residential energy behavior. As output, it generates the following electrical data at various points in the grid:

- Voltage (V) values at each node of the system.
- Current (A) values at each electrical connection.

These results are compared with regulatory limits to ensure the validity of the model. Figure 1 shows the schematic of the microgrid modeled in Simulink, with a detailed view of its structure.

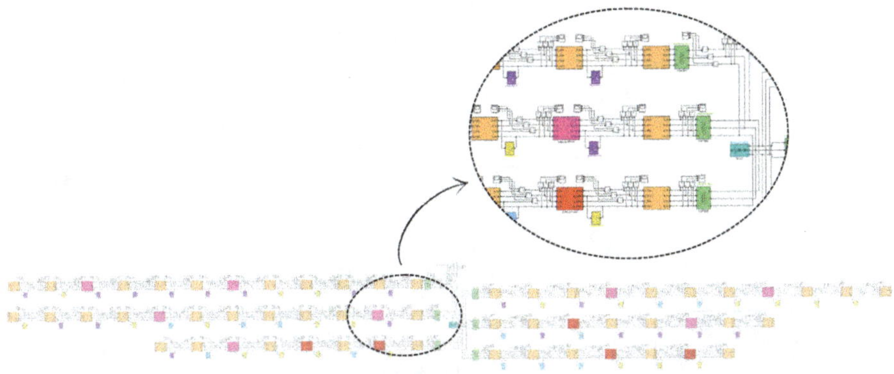

Fig. 1. Simulink Model.

3.2 Machine Leaning

The data generated in the simulation is used to develop a graph-based model, representing the electrical grid as a structured system:

- The nodes represent the system's entities, including households, substations, and the main grid. Each node stores information about energy consumption and production (if applicable).
- The edges represent the electrical connections between the different nodes, incorporating information about the current and voltage at each point in the grid.

This approach enables the application of Graph Neural Networks and other machine learning algorithms to:

1. Predict network current intensity and detect potential overcurrents that could compromise its stability.
2. Optimize microgrid management by identifying strategies to improve efficiency.
3. Detect anomalies in the system, facilitating decision-making.

The model is trained with simulated scenarios to assess its predictive capability and adaptability to different grid configurations (Fig. 2).

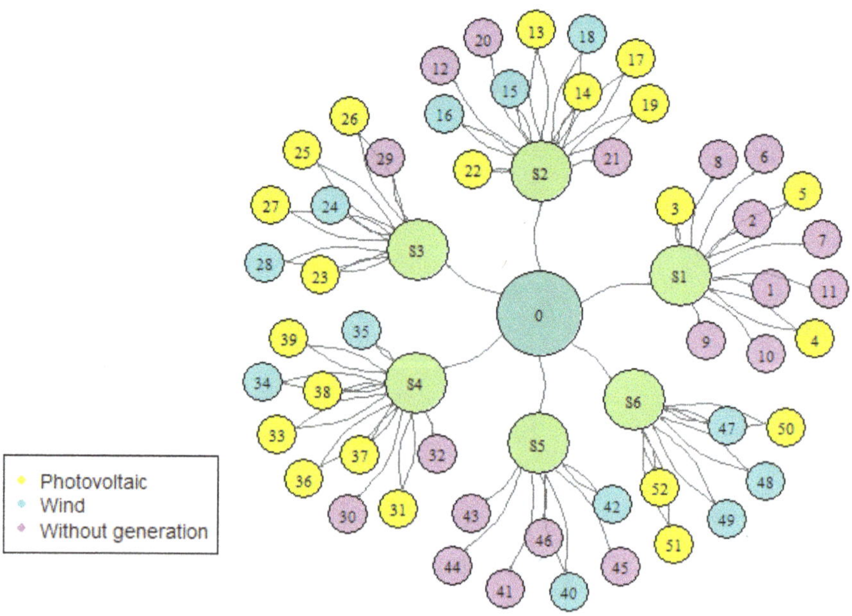

Fig. 2. Graph structure visualization.

4 Results

The objective of this article is to design machine learning systems to predict current intensity one hour in advance to check for overcurrents in the grid. Four different metrics will be calculated to evaluate these predictions:

- **R squared (R^2):** The coefficient of determination, often referred to as R-squared, is the proportion of the total variance in the dependent variable that can be explained by the estimated regression model.

$$R^2 = \frac{\sum_{i=1}^{N} (\hat{y_i} - y_i)^2}{(y_i - \overline{y})^2}$$

- **Mean Absolute Error (MAE):** MAE is used to quantify the prediction accuracy. It measures the average magnitude of the errors in a set of predictions, providing a numerical value that represents how far the predictions are from the actual values.

$$MAE = \frac{\sum_{i=1}^{N} |y_i - \hat{y_i}|}{N}$$

- **Mean Square Error (MSE):** Measures the average of the squares of the errors, i.e. the average squared difference between the estimated values and the actual value.

$$MSE = \frac{\sum_{i=1}^{N} (y_i - \hat{y_i})^2}{N}$$

- **Root mean square error (RMSE):** This is another metric used to measure the accuracy of the prediction.

$$RMSE = \sqrt{\frac{\sum_{i=1}^{N} (y_i - \hat{y_i})^2}{N}}$$

Where $\hat{y_i}$ are the predicted values, y_i the observed values, $\overline{y}$ denotes the mean value, and N is the number of observations.

To make predictions, it is essential to select an appropriate model. Four different classifiers will be tested to assess which one best suits this type of data and to identify the optimal tool for accurately predicting the electrical grid's current intensity and anticipating potential overcurrents.

Figure 3 displays the results graphically, with the x-axis representing the metric values associated with each edge.

The linear regression model (LM) has generally proven to be the most effective, delivering the best results in most predictions. However, its performance declines in the final sections of the line, where the intensity in certain phases is zero due to the absence of consumption or generation in those areas. Analyzing the metrics, it has been observed that, in the case of MAE, the XGBoost and SVM models make better predictions at these endpoints. For MSE, RMSE, and R^2 metrics in these same areas, the SVM model provides the best results.

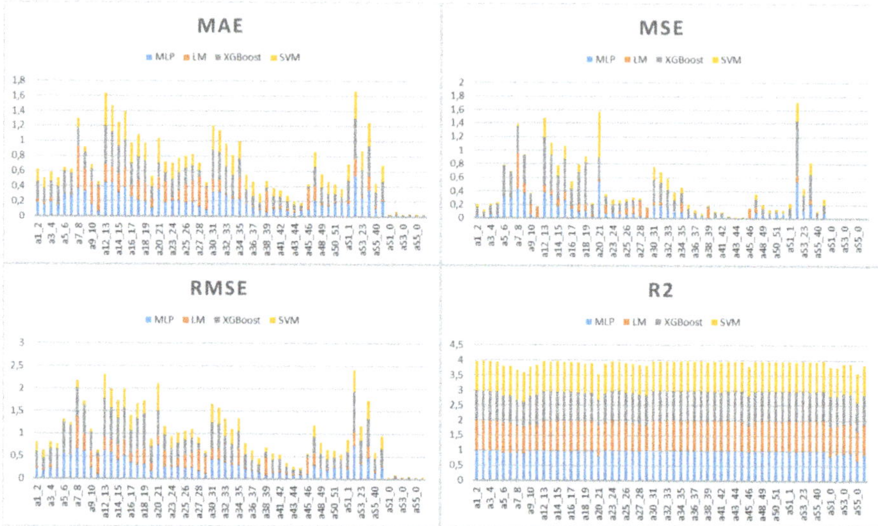

Fig. 3. Results of the classifiers.

Regarding the coefficient of determination (R^2), the obtained values have generally been very satisfactory, indicating a strong correlation between the training data and the predictions made. Furthermore, it has been identified that the edges connecting the network to the substations show highly favorable values across various evaluation metrics. This is likely due to the absence of residential areas at their ends, which reduces variability in energy consumption and injection.

5 Conclusion and Future Work

The prediction of overcurrents in electrical networks is essential for maintaining system stability, preventing failures, and optimizing energy distribution. The application of machine learning models allows for anticipating unexpected current increases, facilitating the implementation of real-time mitigation strategies.

One of the main benefits of this prediction is the protection of electrical infrastructure. When current exceeds nominal limits, system components can overheat, increasing the risk of failures in transformers, circuit breakers, and other critical elements. Early detection of these events enables proactive intervention, preventing costly power outages and reducing system wear.

Furthermore, overcurrent prediction is closely related to the detection of overloads and fluctuations in energy demand. These variations can compromise the quality of electricity supply, causing voltage drops or surges that affect both residential and industrial consumers. Machine learning-based predictive models allow for anticipating these changes and implementing corrective actions, such as load redistribution or the activation of reactive power compensation systems.

Another key aspect is its impact on electrical safety and consumer protection. Overcurrents can unintentionally activate protection devices or, in the worst case, cause fires in installations with inadequate wiring. Early prediction of these events allows for the activation of control mechanisms before the current reaches dangerous levels, minimizing risks for both people and equipment.

The analysis of the study results reveals that the linear regression model (LM) has shown solid overall performance, but its accuracy decreases in the final sections of the line, where the intensity is null in some phases. In these cases, the XGBoost and SVM models have demonstrated better performance, especially in the MAE, MSE, RMSE, and R^2 metrics at the endpoints. This suggests that combining different models could enhance accuracy under various operating conditions.

It has also been identified that the edges connecting the network to substations exhibit favorable values in evaluation metrics, which may be attributed to lower variability in consumption and energy injection in these areas. This finding underscores the importance of considering network structure and demand characteristics when designing more robust prediction models.

In conclusion, the ability to predict overcurrents in electrical networks is a key tool for improving the efficiency, safety, and sustainability of the electrical system. The integration of advanced artificial intelligence models with real-time data enables the development of more resilient systems, capable of adapting to future energy demands and enhancing decision-making in network management.

References

1. Adewuyi, O.B., Krishnamurthy, S.: Appraisal of viable optimization techniques for efficient design and control planning of renewable energy-based microgrid systems. Comput. Electr. Eng. **120**, 109652 (2024)
2. Barja Martínez, S.: Energy management systems for smart homes and local energy communities based on optimization and artificial intelligence techniques (2023)
3. Chen, J., Yu, T., Pan, Z., Zhang, M., Deng, B.: A scalable graph reinforcement learning algorithm based stochastic dynamic dispatch of power system under high penetration of renewable energy. Int. J. Electr. Power Energy Syst. **152**, 109212 (2023)
4. Etxeberria, A., Vechiu, I., Camblong, H., Vinassa, J.M.: Hybrid energy storage systems for renewable energy sources integration in microgrids: a review. In: 2010 Conference Proceedings IPEC, pp. 532–537. IEEE (2010)
5. Luthander, R., Widén, J., Nilsson, D., Palm, J.: Photovoltaic self-consumption in buildings: a review. Appl. Energy **142**, 80–94 (2015)
6. Misic, A.S., Karatas, M., Dasci, A.: Energy storage and transmission line design for an island system with renewable power. Comput. Ind. Eng. 110901 (2025)
7. Ndukwe, C., Iqbal, T.: Sizing and dynamic modelling and simulation of a standalone PV based DC microgrid with battery storage system for a remote community in Nigeria. J. Energy Syst. **3**(2), 67–85 (2019)
8. Porlan-Ferrando, L., Riera-Lorenzo, R., Hernandez-Lecuona, L., Aravena-Cifuentes, A., Nuñez-Gonzalez, J.: A simulation on graph-based learning for energy generation-consumption prediction in energy communities. In: Modelling and Simulation 2024 - 38th Annual European Simulation and Modelling Conference 2024, ESM 2024, pp. 199–204 (2024). https://research.science.eus/documentos/675f91e460134444f311fa8d

9. Soussi, A., Zero, E., Bozzi, A., Sacile, R.: Enhancing energy systems and rural communities through a system of systems approach: a comprehensive review. Energies **17**(19), 4988 (2024)
10. Su, Z., Zheng, G., Hu, M., Kong, L., Wang, G.: Short-term load forecasting of regional integrated energy system based on spatio-temporal convolutional graph neural network. Electr. Power Syst. Res. **232**, 110427 (2024)

Early Data Fusion in Low-Light Vision for YOLO

Paweł Niedziółka(✉)

Wrocław University of Science and Technology, Faculty of Information and Communication Technology, Wybrzeże Stanisława Wyspiańskiego 27, 50-370 Wrocław, Poland
pawel.niedziolka@pwr.edu.pl

Abstract. Object detection presents unique challenges in low-light conditions due to low contrast and limited feature visibility. This work examines the effectiveness of early data fusion for the methods of visible and infrared images within YOLO-based object detection. Several pressing strategies were used, including single-modal training, sequential fine-tuning, and the fused data approach. The YOLO11x model was chosen due to its state-of-the-art performance and adaptability. The experimental evaluation was performed based on the LLVIP database. The results suggest that early fusion, especially direct stacking, can match or cross the baseline model trained only on visual images. On the other hand, specific fusion methods such as Weighted Sum Fusion or the Sequential approach may degrade the performance. This highlights the importance of tailoring fusion strategies for this dataset and motivates further research on hybrid and late fusion techniques in multimodal object detection.

Keywords: Object detection · Data preprocessing · Early data fusion

1 Introduction

Object detection is a crucial and complex task in computer vision. Recently, the field has grown rapidly and the latest advances are perceptible [3]. Reference data sets and evaluation criteria are becoming increasingly structured [8]. It still presents various challenges in broad sections of applications, such as face, defect, text, pedestrian, or remote sensing image detection. The main difficulties in that area are background interference, generalization ability, occlusion problem, multiscale detection, and small object detection [7].

Detection in low-light conditions is a particularly challenging task for applications, and dedicated algorithms have been shown to improve the quality of the results [13]. If available, additional modalities can be used to improve the quality of the prediction. Techniques such as early fusion can be used for this purpose [1].

The main contributions of this work include:

P. Trajdos and R. Burduk (Eds.): CORES 2025, LNNS 1557, pp. 75–82, 2025.
https://doi.org/10.1007/978-3-032-01773-4_8

- Use of the latest available version of the YOLO model for the problem of object detection in low-light conditions employing simple, early data fusion methods.
- Examination of how fine-tuning such a large model affects the quality of the prediction and how the model adapts to these conditions.

2 Experimental Evaluation

The emergence of new, more advanced deep models can raise many questions regarding their effectiveness. The experiments aim to test the model's capabilities and adaptability across low-light conditions. Therefore, the following research questions were posed:

RQ1 - Does using simple algorithms for early data fusion in the low-light vision for huge models (with millions of parameters) give competitive detection performance?

RQ2 - How do the data used to fine-tune the YOLO11x model affect the quality of model classification for multimodal data?

2.1 Dataset

The dataset used for the experiments is LLVIP [4]. It is multimodal and contains visible (RGB) and infrared (IR) images for low-light vision problems. The data were first pre-processed accordingly. It was not necessary to standardize the dimensions of the images, since both types had a resolution of 1280×1024. The subsets were divided and extracted in the appropriate proportions: training (70%), validation (15%), and test (15%). Annotations were also converted from the Pascal Visual Object Classes type to one that matches the YOLO architecture.

Data fusion techniques provide some challenges at the data level, such as acquisition and observation or various types of uncertainty [6]. Due to various factors, not all data entries have the same degree of confidence, reliability, or information quality [12].

It was determined at the early stage that employing straightforward methods would be the optimal approach to fuse and consolidate the two modalities. The visualization of these methods is visible in Fig. 1.

Direct Stacking Fusion (DSF) involves concatenating standard RGB and IR images along a new channel dimension, resulting in a four-channel input. This method preserves the complete information from both sources. However, it may require additional requirements for compatibility with standard deep-learning models that handle only three-dimensional channel inputs.

Weighted Sum Fusion (WSF) integrates the RGB and IR images by blending their pixel values from defined weights. IR images often have intensity values that differ from RGB ones, so the image must be normalized to maintain visual coherence and properly convert to match the dimensions. The final fused image is computed as a weighted sum using Eq. 1. This method amplifies complementary features from both modalities, potentially improving overall information quality.

Fig. 1. Visualization of the data fusion techniques.

$$I_f(x, y) = \alpha \cdot I_{RGB}(x, y) + \beta \cdot I_{CIR}(x, y) \quad (1)$$

where:

$I_f(x, y)$ - the resulting fused image
$I_{RGB}(x, y)$ - the input RGB image
$I_{CIR}(x, y)$ - the infrared converted to multichannel image by replicating channels accordingly: $[I_{IR}(x, y), I_{IR}(x, y), I_{IR}(x, y)]$
$I_{IR}(x, y)$ - the input IR image
α, β - the weighting coefficients

The label correlogram offers valuable insight into the relationships between various label dimensions in the dataset. Highlighting patterns and correlations can improve understanding of the distribution of object annotations, providing helpful context for data analysis and model development. Looking at the correlogram (Fig. 2), the width and height histograms indicate that the LLVIP dataset appears well distributed regarding object positions but may be biased towards smaller ones. Additionally, a positive correlation between width and height indicates the presence of objects with consistent proportions.

2.2 Models

In the past, models with Region-based Convolutional Neural Networks architecture reigned in the object detection models [14]. With exploration and continued development came single-stage models that simultaneously feature prediction of all the bounding boxes and have feed-forward architecture such as Single Shot

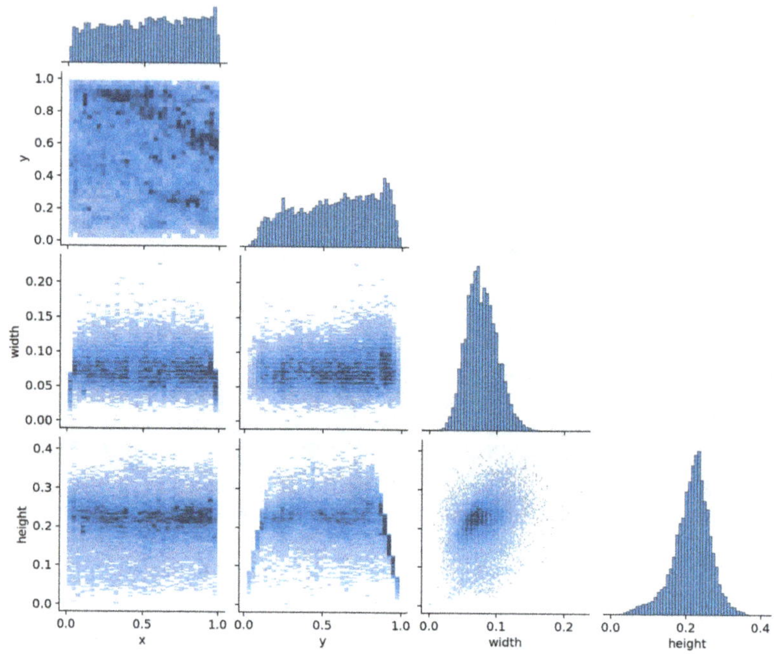

Fig. 2. LLVIP labels correlogram.

multibox Detector [9] or YOLO [10]. YOLO is a computer vision model for object detection, classification, and segmentation. It performs significantly better than its counterparts in the two-stage object detector, excelling in detection accuracy and inference speed [2].

The YOLO model was chosen for experimentation due to several factors, mainly for its popularity and continuous improvement in development. In addition, its practical advantages, architecture that avoids the need for complex region proposal mechanisms, and simple public API contributed to the choice of this model. At first, YOLO7 was used, but after some architectural issues, it was decided to change the version to YOLO11 with the designation 'x' [5]. This is the most extensive type of this model, with 56.9M parameters and 194.9B FLOPs. It was the best possible model for experiments, while another version - YOLO12 [11] - was recently released – February 18th, 2025. This only further illustrates how rapidly the field is developing.

2.3 Experimental Protocol

The libraries used in the experiments are PyTorch 2.1.0, Python 3.10, Cuda 11.8.0. The training process took place on Ubuntu 22.04 with a dedicated Nvidia A40 GPU (48GB RAM). This allowed one to choose a larger batch size and process data simultaneously. The fine-tuning of the model was done for 100 epochs with a batch size of 32, and the sizes of the images passed to the model were set to 640×640px.

2.4 Reproducibility

YOLO models and the LLVIP dataset are publicly available. The code for the preparation of the dataset (described in Sect. 2.1) is publicly available on GitHub[1]. As the dataset is split randomly during preparation, the prepared splits are available for download via Google Drive to ensure reliability. The URLs are located in the llvip directory in the README file.

2.5 Results

Firstly, the model was fine-tuned in several distinct configurations to evaluate performance across various data modalities:

- **Visible-Only** The model was trained exclusively on the visible spectrum images.
- **Infrared-Only** The model was trained solely using infrared images.
- **Sequential** In this approach, the model was first pre-trained on visible images and subsequently fine-tuned using infrared data.
- **Direct Stacking Fusion (DSF)** The model was trained on a fused dataset generated through direct stacking (described in Sect. 2.1) of visible and infrared images.
- **Weighted Sum Fusion (WSF)** The final configuration involved pretraining the model on a fused dataset created using the weighted sum fusion method (described in Sect. 2.1) with weights $\alpha = 0.7$ and $\beta = 0.3$. These values were selected heuristically based on preliminary visual inspection aiming to preserve the dominant characteristics of the RGB image while incorporating complementary information from the infrared modality.

Out of the box, the images are resized in unpredictable ways during the pre-training process to avoid overfitting for similar photos. The example of the learning batch is visible in Fig. 3.

The Yolo11 model with the standard weights was fine-tuned in every configuration listed in Sect. 2.5. Later, these models were used on a separate validation subset for the object detection task.

Figure 4 presents the training and validating curves for box loss, classification loss (CSL), and distribution focal loss (DFL) in multiple pre-training strategies applied to the YOLO11x model. The YOLO11 baseline, trained on the COCO dataset without specific tuning on LLVIP, performs worst in convergence, especially in CSL and DFL, highlighting the importance of domain adaptation. Models pre-trained on visible images and with DSF show the most stable and the lowest loss values in training and validation sets. This shows

[1] https://github.com/KoEj/yolov7/tree/yoloV7_LLVIP.

Fig. 3. Vizualization of the training batch for the YOLO.

that these strategies are more effective in preserving critical features. WSF and Sequential approaches show moderate performance, indicating a potential mismatch or inefficient feature transfer. The model trained only on infrared data shows an inferior generalization and should not be used in isolation.

Table 1. Models fine-tuned, and tested on visible images

	YOLO11X	Visible	Infrared	Sequential	DSF	WSF
mAP50-95	0.3462	**0.5864**	0.0117	0.0556	**0.5873**	0.2274
mAP50	0.6860	**0.9509**	0.0297	0.1418	**0.9492**	0.5006
mAP75	0.3034	**0.6349**	0.0071	0.0345	**0.6335**	0.1755

Analyzing the results obtained from the experiment shown in Table 1, the infrared modality contains features that differ from those in the visible spectrum. As a result, infrared data alone may not be sufficient for effective object detection when used in seclusion. Sequential and WSF approaches also do not fit this kind of data. One of the fusion methods, DSF, has similar results to those obtained from the model on which the collection was tested.

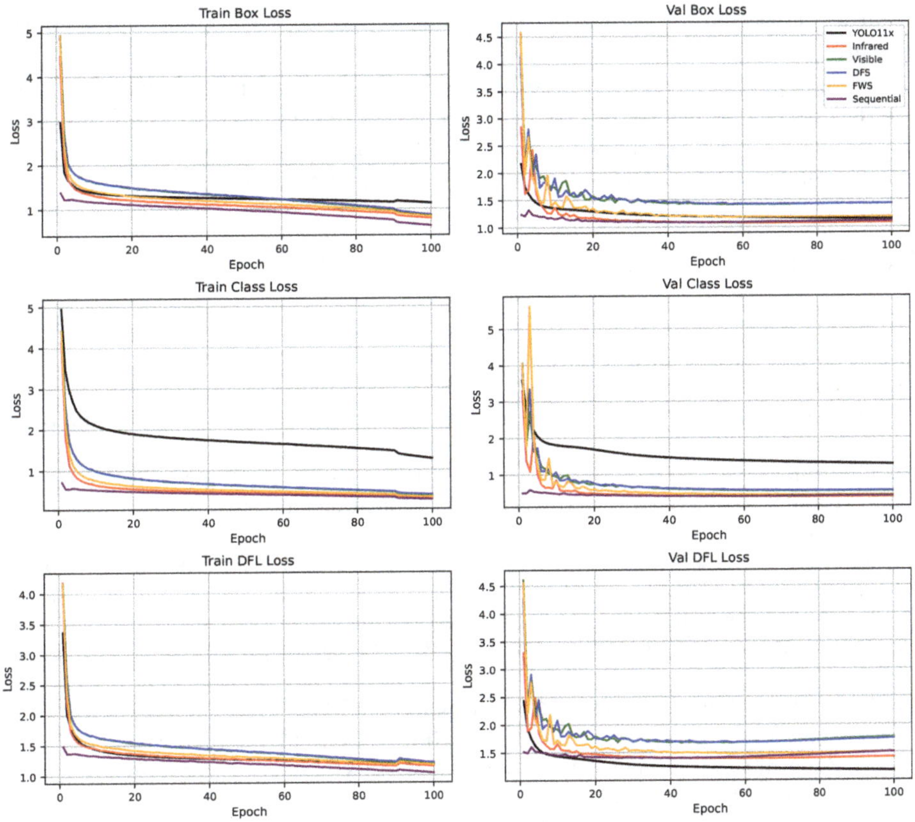

Fig. 4. Metrics for training and validation curves.

2.6 Lesson Learned

The experiments show that simple algorithms can positively impact the prediction quality of large models. However, no significantly better results were obtained, so this cannot be definitively confirmed. Further research would be necessary using a wider range of simple early-fusion algorithms. Additionally, results indicate that choosing the wrong modality can significantly drop prediction accuracy; therefore, this process should be approached carefully.

3 Conclusion

Despite a model consisting of millions of parameters, not all data fusion methods prove universally successful. The effectiveness of each approach must be carefully evaluated and validated for each specific dataset to ensure optimal performance. Early fusion techniques look promising for YOLO architecture. Investigating whether more advanced or alternative fusion strategies can improve detection accuracy is necessary. However, the process will be both time and resource-consuming.

More experiments should be conducted to better understand the accuracy of the detection. Future work includes research on middle, hybrid, and late fusion techniques, as they can offer improvements and insights. However, it can be seen that fusion methods (or lack thereof) matter for detection outcomes.

References

1. Barnum, G., Talukder, S., Yue, Y.: On the benefits of early fusion in multimodal representation learning. arXiv preprint arXiv:2011.07191 (2020)
2. Diwan, T., Anirudh, G., Tembhurne, J.V.: Object detection using yolo: challenges, architectural successors, datasets and applications. multimedia Tools Appl. **82**(6), 9243–9275 (2023)
3. Gallagher, J.E., Oughton, E.J.: Surveying you only look once (YOLO) multispectral object detection advancements, applications and challenges. IEEE Access (2025)
4. Jia, X., Zhu, C., Li, M., Tang, W., Zhou, W.: LLVIP: a visible-infrared paired dataset for low-light vision. In: Proceedings of the IEEE/CVF International Conference on Computer Vision, pp. 3496–3504 (2021)
5. Jocher, G., Qiu, J.: Ultralytics YOLO11 (2024). https://github.com/ultralytics/ultralytics
6. Lahat, D., Adali, T., Jutten, C.: Multimodal data fusion: an overview of methods, challenges, and prospects. Proc. IEEE **103**(9), 1449–1477 (2015)
7. Li, Z., et al.: Development and challenges of object detection: a survey. Neurocomputing **598**, 128102 (2024)
8. Liu, L., et al.: Deep learning for generic object detection: a survey. Int. J. Comput. Vision **128**, 261–318 (2020)
9. Liu, W., et al.: SSD: Single Shot MultiBox Detector. In: Leibe, B., Matas, J., Sebe, N., Welling, M. (eds.) ECCV 2016. LNCS, vol. 9905, pp. 21–37. Springer, Cham (2016). https://doi.org/10.1007/978-3-319-46448-0_2
10. Redmon, J., Divvala, S., Girshick, R., Farhadi, A.: You only look once: unified, real-time object detection. In: Proceedings of the IEEE Conference on Computer Vision and Pattern Recognition, pp. 779–788 (2016)
11. Tian, Y., Ye, Q., Doermann, D.: YOLOv12: Attention-centric real-time object detectors. arXiv preprint arXiv:2502.12524 (2025)
12. Turk, M.: Multimodal interaction: a review. Pattern Recogn. Lett. **36**, 189–195 (2014)
13. Xiao, Y., Jiang, A., Ye, J., Wang, M.W.: Making of night vision: object detection under low-illumination. IEEE Access **8**, 123075–123086 (2020). https://doi.org/10.1109/ACCESS.2020.3007610
14. Zou, Z., Chen, K., Shi, Z., Guo, Y., Ye, J.: Object detection in 20 years: a survey. Proc. IEEE **111**(3), 257–276 (2023). https://doi.org/10.1109/JPROC.2023.3238524

CorroSeg: U-Net Based Corrosion Segmentation in Steel Pipelines Using Ultrasonic Images

Mounir Salhi(✉) and Moulay A. Akhloufi

Perception, Robotics, and Intelligent Machines Lab (PRIME), Department of Computer Science, Université de Moncton, Moncton, Canada
{ems9346,moulay.akhloufi}@umoncton.ca

Abstract. Corrosion in steel pipelines is a critical issue in the oil and gas industry. Early corrosion detection can limit the environmental damage and financial impact. To address this challenge, a new method was developed to segment corrosion defects from ultrasonic images representing variations in pipeline thickness. Different neural network architectures including 5-layers CNN, Mini U-Net, Deep U-Net and ResNet50 were evaluated. Among them, the Deep U-Net demonstrated superior performance with a Dice score of 0.845 and an IoU of 0.732 on validation data, with an IoU of 0.6410 using test data, outperforming existing methods. These results show the potential of Deep U-Net for segmenting corrosion in steel pipelines. These improvements may significantly facilitate the pipeline inspection process by reducing the required manual effort.

Keywords: Corrosion detection · Deep learning · Ultrasonic imaging · Image segmentation · U-Net

1 Introduction

Corrosion in steel pipelines is one of the most critical challenges in the oil and gas industry, accounting for more than 25% of all well failures [8]. In addition to the environmental risks, corrosion costs are estimated at US$1,372B annually [11]. Preventing these consequences requires early and precise detection of structural defects to react before critical failures occur. Ultrasonic images are used to analyze pipelines thickness variations and identify corrosion patterns. However, the manual segmentation of corrosion in these ultrasonic images is very time-consuming due to the large amount of data, which makes the process prone to human errors [9]. Several existing approaches have been proposed to automate corrosion segmentation, such as the CNN-LSTM network [7], RustSEG [3], and Mask R-CNN [2], which have proven their effectiveness for segmenting corrosion defects. However, these approaches mainly rely on classic RGB images of the tubes, limiting their effectiveness in segmenting surface defects. Non-destructive inspection methods such as ultrasonic, electromagnetic, and mechanical tools are

P. Trajdos and R. Burduk (Eds.): CORES 2025, LNNS 1557, pp. 83–92, 2025.
https://doi.org/10.1007/978-3-032-01773-4_9

used to inspect and verify the condition of steel pipelines and detect corrosion. These tools provide detailed maps of variations in tube thickness. These data are used to identify defects such as local corrosion and axial grooves. Due to telemetry, processing errors can occur, corrupting some data points on the maps. To overcome these limitations, the U-Net architecture introduced by Ronneberger et al. (2015) [12] has demonstrated its effectiveness on ultrasound image segmentation in the medical field [1,4,14], suggesting its potential for ultrasonic corrosion segmentation due to the need for precise edge detection and preservation of spatial details. In this work, we explored and evaluated various deep learning architectures for corrosion segmentation in ultrasonic images (CNN, Mini U-Net, ResNet50 [6], Deep U-Net). We implemented a combined loss function to handle severe class imbalance. We demonstrate the U-Net based architecture's efficiency in segmenting both inner and outer corrosion defects from ultrasonic images.

RQ1: Can a Deep U-Net outperform standard CNNs, lightweight U-Nets, and transfer learning models on ultrasonic corrosion segmentation? RQ2: What's the best trade-off between U-Net model depth and segmentation performance?

The remainder of the paper is structured as follows: Sect. 2 reviews the related work. Section 3 presents the approaches and methodologies used for this study. Section 4 provides a summary and analysis of the experimental results. Section 5 concludes and discusses potential future work.

2 Related Work

This section reviews modern approaches based on artificial intelligence and ultrasonic wave technologies for corrosion detection. Lyu et al. [10] took the first step in capturing the use of guided ultrasonic waves in research on corrosion detection in pipelines. They discuss Lamb waves and horizontally traveling shear waves for pipeline inspection. These waves, which propagate over long distances, also penetrate materials, making them reliable for identifying internal defects such as corrosion. However, the approach needs selecting the correct wave mode and frequency, which are crucial for segmenting corrosion from grayscale ultrasonic images. Huang et al. [7] proposed a hybrid model combining a convolutional neural network (CNN) with a long short-term memory (LSTM) network to process ultrasonic signals. In this approach, CNN extracts spatial features and LSTM captures temporal links. Burton et al. [3] developed RustSEG model, which uses AI to segment corroded surfaces from visual images. One of the main strengths of this method is that it operates without requiring pixel-level annotations (masks), a major constraint in many deep-learning approaches. RustSEG relies on a convolutional neural network and class activation maps (CAM). The model initially classifies images and generates heatmaps highlighting corrosion areas, refined by a conditional random field to segment corrosion precisely. Ameli et al. [2] used Mask RCNN and YOLOv8, two architectures that perform segmentation and object detection respectively. These models effectively segment irregular corrosion shapes, creating precise binary masks even in noisy conditions. Ronneberger et al. [12] developed the U-Net architecture, designed for biomedical image segmentation, especially on grayscale images such as electron microscopy and ultrasound data. U-Net is based on an encoder-decoder structure with skip connections between the corresponding layers, allowing it to capture both global context and fine details. The model has shown excellent results on tasks with limited number of annotated images. The U-Net architecture is particularly suitable for tasks such as ultrasonic corrosion detection due to its ability to segment grayscale images with high precision, even under noisy and low-contrast conditions.

3 Methods

The U-Net architecture [12] was chosen due to its capacity to accurately learn the patch characteristics and perform more accurate segmentation. We adapted this model by increasing its depth to five encoder and decoder blocks instead of four. We refer to this new version as Deep U-Net in this paper. We resized the images to 128×128 pixels, enabling it to exploit its full depth of 5 pooling levels for efficient feature extraction. The architecture of the chosen model (Fig. 1) has five level encoder-decoder structure. Each encoder block has two convolutional layers with a progressive increase in filters (64, 128, 256, 512, 1024), and a pooling layer. The bottleneck part consists of two convolutional layers with 1024 filters each, extracting the most abstract features before the decoder part. The decoder contains an UpSampling to progressively reconstruct the size of the input image, a concatenation with features extracted with the encoder, and two convolution layers with decreasing filters (512, 256, 128, 64, 64) to refine the segmentation. Finally, the output consists of a single 1×1 convolutional layer with a sigmoid activation function that generates probabilities indicating the presence or not of corrosion pixel by pixel. The choice of an architecture with 5 levels in the encoder and decoder is justified by the need to capture fine details and larger image contexts. Increasing filters in the encoder is for the extraction of more complex features at each level while decreasing filters in the decoder helps to reconstruct the corrosion masks. The central bottleneck part with its two convolutional layers of 1024 filters is for the efficient processing of information.

Other architecture options were explored for their ability to handle single-channel ultrasonic images (grayscale) and effectively segment corrosion. 5-layer CNN is the simplest model tested and serves as the baseline architecture [5], it consists of five convolutional layers without pooling, each followed by a ReLU activation. The Mini U-Net is a lighter and shallow version of U-Net [12],composed of two encoding blocks and two decoding blocks. Pretrained ResNet50 was also evaluated, using ResNet50 [6] as a pretrained encoder on ImageNet to extract deep features, and a decoder composed of transposed and standard convolutional layers. Finally, Deeper U-Net is the deepest version of the U-Net architecture [12] tested with 7 encoding blocks and 7 decoding blocks.

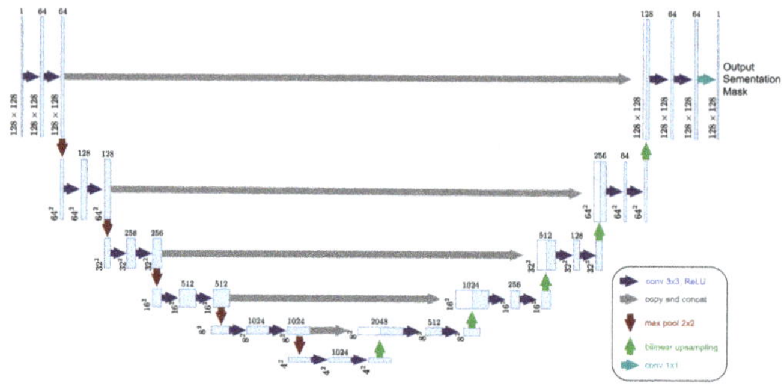

Fig. 1. Overview of the Deep U-Net architecture (adapted from the original U-Net [12]) used in this study. Blue boxes represent feature maps, with the number of channels on the top and dimensions at the lower left edge of the box. The arrows represent different operations.

3.1 Combined Loss Function

To handle the class imbalance and improve the accuracy of predictions, a combined loss function has been implemented as defined in Eq. 1 with $\alpha = 2.0$ and $\beta = 1.5$.

$$\mathcal{L}_{\text{combined}} = \alpha \cdot \mathcal{L}_{\text{BCE-w}} + \beta \cdot \mathcal{L}_{\text{Dice}} \tag{1}$$

$$\mathcal{L}_{\text{BCE-w}} = -\frac{1}{N}\sum_{i=1}^{N} w_{y_{\text{true},i}} \left[y_{\text{true},i}\log(y_{\text{pred},i}) + (1 - y_{\text{true},i})\log(1 - y_{\text{pred},i})\right] \tag{2}$$

$$\mathcal{L}_{\text{Dice}} = 1 - \frac{2\sum_{i=1}^{N} y_{\text{true},i} \cdot y_{\text{pred},i} + \epsilon}{\sum_{i=1}^{N} y_{\text{true},i} + \sum_{i=1}^{N} y_{\text{pred},i} + \epsilon} \tag{3}$$

The Weighted Binary Cross-Entropy loss (Eq. 2) penalizes the difference between predictions and ground truth, assigning weights $w_0 = 0.5$ to background pixels and $w_1 = 1.5$ to corrosion pixels, to handle class imbalance.

The Dice loss defined in Eq. 3, optimizes the Dice coefficient, which correlates positively with the Intersection over Union (IoU), aiming to maximize the overlap between predicted masks and ground truth. In a context where corrosion areas are rare, this loss helps better capture these small regions. Dice-based loss functions have shown their effectiveness on highly unbalanced classes segmentation [13].

3.2 Thresholding

After training, we identify the optimal binary threshold that maximizes the Intersection over Union (IoU) score between predicted probabilities Y_{pred} and ground truth masks Y_{true}. For a set of thresholds $t \in [0, 1]$, the binarized prediction is defined in Eq. 4, where $\mathbf{1}_{\{\cdot\}}$ is the thresholding function that assigns 1 to pixels above the threshold t or 0 otherwise. The optimal threshold t^*, as defined in Eq. 6, is found using a grid search with 10000 thresholds sampled in [0,1] to maximize the IoU metric defined in Eq. 5. This threshold t^* is then applied to binarize all predictions during evaluation.

$$\hat{Y}_t = \mathbf{1}_{\{Y_{\text{pred}} \geq t\}} \tag{4}$$

$$\text{IoU}(\hat{Y}_t, Y_{\text{true}}) = \frac{|\hat{Y}_t \cap Y_{\text{true}}|}{|\hat{Y}_t \cup Y_{\text{true}}|} \tag{5}$$

$$t^* = \underset{t \in [0,1]}{\arg\max}\, \text{IoU}(\hat{Y}_t, Y_{\text{true}}) \tag{6}$$

4 Results

In this section, we present the experimental results and compare the performance of the Deep U-Net against other tested architectures.

4.1 Corrosion Dataset

The data used in this work was sourced from [5]. The dataset includes ultrasonic grayscale images of 20 wells. Each image in the dataset is associated with a binary mask that has the same dimensions and indicates the presence (1) or absence (0) of corrosion. The total number of patches is 9674, each of size 36×36 pixels, including some corrupted patches with NaN values. We used only the valid patches (9332 patches) and their corresponding masks. The distribution of corrosion and non-corrosion classes is unbalanced with 12.97 times more pixels indicating the absence of corrosion. To summarize, among all the pixels present in the masks, 92.85% are non-corrosion (0) and 7.15% are corrosion (1). Data were randomly split into 8398 training patches (90%) and 934 validation patches (10%). Results in Tables 1 and 2 are averaged over 5-fold cross-validation. The test set (without masks) was provided separately and evaluated externally via the official challenge platform[5]. The original patches of size 36×36 pixels have a resolution that limits the depth of the models that can be used, as architectures like U-Net require a progressive reduction in image size through pooling layers. Resizing the patches from 36×36 to 128×128 pixels using TensorFlow's function with bilinear interpolation allowed for an increase in resolution while preserving the main patterns and structures of the patches and their corrosion masks. This gives the model more pixels per corrosion area, thereby improving its ability to learn edges and small regions.

4.2 Results Analysis

Models were trained using Adam optimizer with an initial learning rate of 0.001, halved every 5 epochs if the validation loss does not improve. A batch size of

64 was chosen to find a good compromise between model stability by reducing gradient fluctuations and GPU memory constraints. The loss function used is the combined loss, as defined in Eq. 1 in Sect. 3, which allowed for improved results. Augmentation techniques used are image rotation within an interval of [-5,+5]°, vertical shift of 10%, and image flipping to simulate realistic variations in the data collected. The number of epochs for the initial training was 50 epochs to evaluate the different architectures, then an extended training of 200 epochs for selected models.

The performances of the different models explored were compared using the Dice coefficient and IoU score as evaluation metrics. Table 1 shows the results obtained after 50 training epochs on the validation dataset. These scores allow us to compare between models performance on early-stage training and to select the most promising models.

Table 1. Performance on Validation dataset after 50 training epochs

Model	Val Dice coefficient	Val IoU
CNN	0.636	0.466
Mini U-Net (2 layers)	0.780	0.631
Transfer learning (ResNet50)	0.132	0.071
Deep U-Net (5 layers)	0.819	0.693
Deeper U-Net (7 layers)	0.822	0.698

For the best-performing and promising models, training was extended to 200 epochs. Table 2 shows their final score on the validation dataset:

Table 2. Performance on validation dataset after 200 training epochs

Model	Val Dice coefficient	Val IoU
Mini U-Net (2 layers)	0.791	0.654
Deep U-Net (5 layers)	**0.845**	**0.732**
Deeper U-Net (7 layers)	0.840	0.725

The CNN model obtained a validation Dice coefficient of 0.635 after 50 epochs, no further improvements were achieved.

Mini U-Net achieved a validation Dice coefficient of 0.774 after 50 training epochs, which improved to 0.778 after 200 training epochs. However, its segmentation and feature extraction capabilities are limited by its depth, and its ability to generalize is particularly constrained in complex areas with irregular corrosion boundaries. Figure 2a illustrates the learning curves of the Mini U-Net model. The curves show a consistent decrease in training and validation loss, indicating stable convergence.

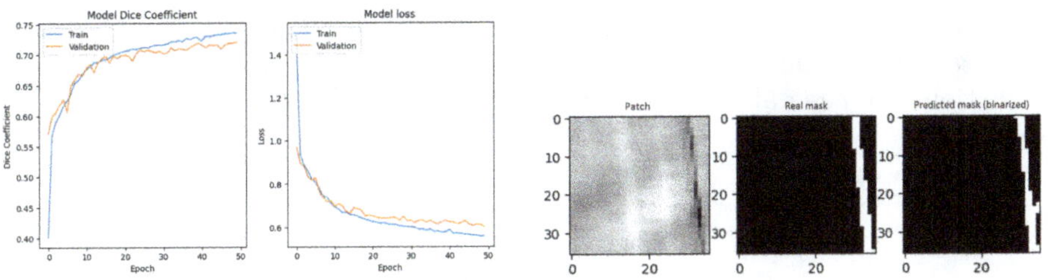

(a) Learning curves of the Mini U-Net (b) Example of Mini U-Net prediction

Fig. 2. Mini U-Net model training and prediction.

An example of prediction is presented in Fig. 2b from the Mini U-Net, which produced segmentation with improved delineation of the edges. However, the model struggles to detect small areas of corrosion.

ResNet50 [6] showed high variance across runs, although an earlier single run achieved a Dice of 0.623, the average performance across 5 folds was substantially lower (0.132), reflecting its instability on this task. Figure 3a illustrates the model's learning curves, showing significant fluctuations despite fine-tuning. Figure 3b shows an example of prediction from the ResNet50 model. The model still struggles with the edges of corrosion areas and small zones, tending to over-segment and under-segment.

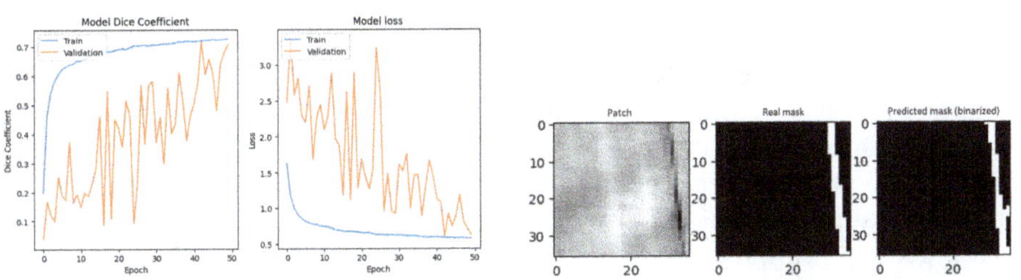

(a) Learning curves of the ResNet50 (b) Example of ResNet50 prediction

Fig. 3. ResNet50 model training and prediction.

Deeper U-Net model achieved a validation Dice coefficient of 0.810 after 50 epochs, and 0.827 after 200 epochs. Deeper U-Net slightly outperformed Deep U-Net after 50 training epochs, but Deep U-Net performed better after 200 epochs of training. Figure 4a shows the learning curves of the Deeper U-Net model over 200 epochs of training. The convergence is slower than that of Deep U-Net due to the complexity of the model. Nevertheless, stability is achieved after approximately 100 training epochs. The gap between the training and validation losses remains small, indicating a good ability of the model to generalize. Figure 4b shows an example of a prediction made by the Deeper U-Net model, the segmentation effectively captures the edges. Sometimes the model tends to over-segment by adding corrosion areas where there are none, but the model's predictions are generally coherent and accurate.

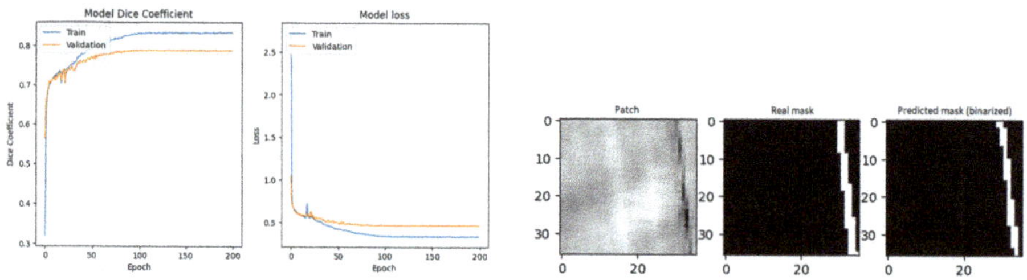

(a) Learning curves of the Deeper U-Net (b) Example of Deeper U-Net prediction

Fig. 4. Deeper U-Net model training and prediction.

Deep U-Net Among the tested architectures, Deep U-Net achieved the best performance in this segmentation task. The Deep U-Net achieved a validation Dice coefficient of 0.805 after 50 training epochs, which improved to 0.845 after 200 epochs. It achieved a validation IoU score of 0.732. Its ability to capture details and features surpasses that of all other tested models, even though it is less complex than Deeper U-Net. In Fig. 5a, the learning curves of the Deep U-Net model over 200 epochs show a quick convergence from the early epochs. After 100 epochs, stability is achieved, with the training and validation loss well-aligned, indicating the absence of overfitting. The use of data augmentation techniques, such as rotations and shifts, has enhanced the model's generalization capability. Deep U-Net maintains a steady progression without major oscillations, highlighting effective adaptation. An example of Deep U-Net prediction after 200 training epochs is illustrated in Fig. 5b, the predictions are accurate with a good capture of the edges of corrosion areas. The model has learned to segment small corrosion areas well and to better preserve the edges of these areas.

4.3 Result on the Test Set

Using the Deep U-Net model trained for 200 epochs, a prediction on the test data was made and then binarized using the optimal threshold for this model. These predictions were submitted for evaluation to the challengedata website [5]. The Deep U-Net achieved an IoU score of **0.6410** on the test set, outperforming both the benchmark score of 0.4798 and Deeper U-Net with an IoU score of 0.6302, as reported in Table 3. These results demonstrate that the Deep U-Net model offers the most favorable trade-off between segmentation accuracy and architectural complexity, while demonstrating strong generalization capabilities for this task. All performance scores were computed after binarizing the predictions using the

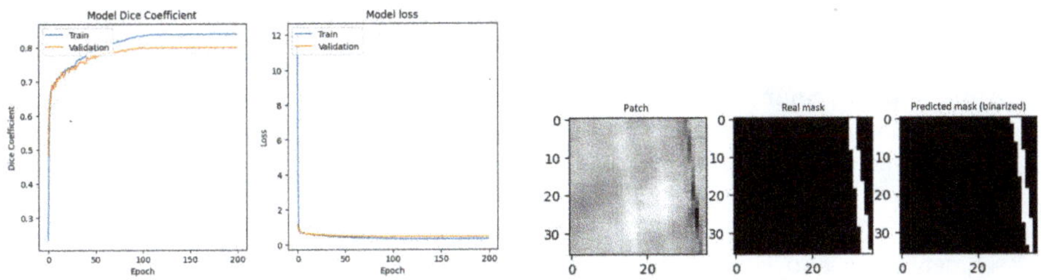

(a) Learning curves of the Deep U-Net (b) Example of Deep U-Net prediction

Fig. 5. Deep U-Net model training and prediction.

optimal threshold determined on the validation set according to the procedure described in Sect. 3.

Table 3. Test IoU scores

Model	Test IoU
Deep U-Net (5 layers)	**0.6410**
Deeper U-Net (7 layers)	0.6302
Benchmark	0.4798

5 Conclusion

This study shows the effectiveness of a Deep U-Net architecture for accurate segmentation of corrosion defects in ultrasonic images of steel pipelines. Patch resizing to 128×128 and applying data augmentation improved the model's ability to delineate the corrosion boundaries. The combined loss function helped to handle the class imbalance and improved the segmentation performance. The Deep U-Net model achieved the highest test IoU of 0.6410, demonstrating its ability to generalize effectively.

However, some challenges remain. The initial size of the patches limited feature extraction, resizing improved accuracy, but resizing the predictions to the original resolution slightly decreased precision. Although the combined loss improved the model's accuracy, small corrosion areas still tend to be under-segmented. Comparison with Deeper U-Net indicates that there is no additional performance gain from further depth.

Future work could explore more data augmentation strategies (e.g., elastic deformation, Gaussian noise), advanced preprocessing to handle partially corrupted data, and alternative architectures such as Transformers. Additionally, techniques such as Focal Loss may improve performance by giving more importance to difficult examples such as small areas of corrosion.

References

1. Almajalid, R., Shan, J., Du, Y., Zhang, M.: Development of a deep-learning-based method for breast ultrasound image segmentation. In: 2018 17th IEEE International Conference on Machine Learning and Applications (ICMLA), pp. 1103–1108 (2018)
2. Ameli, Z., Nesheli, S.J., Landis, E.N.: Deep learning-based steel bridge corrosion segmentation and condition rating using mask RCNN and YOLOv8. Infrastructures **9**(1) (2024). https://doi.org/10.3390/infrastructures9010003
3. Burton, B., Nash, W.T., Birbilis, N.: RustSEG – Automated segmentation of corrosion using deep learning. arXiv:2205.05426 (2022)
4. Cheng, D., Lam, E.Y.: Transfer learning U-Net deep learning for lung ultrasound segmentation. arXiv:2110.02196 (2021)
5. CorroSeg: challenge data. https://challengedata.ens.fr/participants/challenges/144/. Accessed October 2024
6. He, K., Zhang, X., Ren, S., Sun, J.: Deep Residual learning for image recognition. In: Proceedings of the IEEE Conference on Computer Vision and Pattern Recognition (CVPR) (2016)
7. Huang, L., Hong, X., Yang, Z., Liu, Y., Zhang, B.: CNN-LSTM network-based damage detection approach for copper pipeline using laser ultrasonic scanning. Ultrasonics **121**, 106685 (2022). https://doi.org/10.1016/j.ultras.2022.106685
8. Kermani, M.B., Harrop, D.: The impact of corrosion on the oil and gas industry. SPE Prod. Facil. **11**(03), 186–190 (1996). https://doi.org/10.2118/29784-PA
9. Khayatazad, M., De Pue, L., De Waele, W.: Detection of corrosion on steel structures using automated image processing. Dev. Built Environ. **3**, 100022 (2020). https://doi.org/10.1016/j.dibe.2020.100022
10. Lyu, F., Zhou, X., Ding, Z., Qiao, X., Song, D.: Application research of ultrasonic-guided wave technology in pipeline corrosion defect detection: a review. Coatings **14**(3) (2024). https://doi.org/10.3390/coatings14030358
11. Perez, T.E.: Corrosion in the oil and gas industry: an increasing challenge for materials. JOM **65**(8), 1033–1042 (2013). https://doi.org/10.1007/s11837-013-0675-3
12. Ronneberger, O., Fischer, P., Brox, T.: U-Net: convolutional networks for biomedical image segmentation. In: N. Navab, J. Hornegger, W.M. Wells, A.F. Frangi (eds.) Medical Image Computing and Computer-Assisted Intervention – MICCAI 2015, pp. 234–241. Springer, Cham (2015)
13. Sudre, C.H., Li, W., Vercauteren, T., Ourselin, S., Jorge Cardoso, M.: Generalised dice overlap as a deep learning loss function for highly unbalanced segmentations. In: Deep Learning in Medical Image Analysis and Multimodal Learning for Clinical Decision Support, pp. 240–248 (2017)
14. Wang, Z.: Deep learning in medical ultrasound image segmentation: a review. arXiv:2002.07703 (2021)

Analysis of Selected Deep Active Learning Algorithms

Mariusz Topolski[1(✉)], Katarzyna Topolska[1], Jan Wasilewski[2], and Michał Woźniak[1]

[1] Wroclaw University of Science and Technology, Wrocław, Poland
{mariusz.topolski,katarzyna.topolska,michal.wozniak}@pwr.edu.pl
[2] Rochester Institute of Technology, Rochester, NY, USA
jw7630@g.rit.edu

Abstract. Access to correct labels is crucial from the point of view of classifier learning. Unfortunately, the labeling process involves a high cost in many cases, so methods based only on partially labeled data are becoming increasingly popular. This paper presents selected methods of data labeling using active learning. Experimental studies conducted on four benchmark databases confirm the usefulness of the methods chosen and their superiority over object labeling using a random approach.

Keywords: Machine learning · Pattern classification · Active learning

1 Introduction

The increasing utilization of knowledge characterizes the modern economy as a crucial resource and a source of competitive advantage. Large volumes of data and the constantly developing industry make various data collection and processing methods evolve rapidly. With technological progress, the demand for complex classifiers is growing exponentially, requiring the necessary data. One of the problems is data acquisition cost [15], especially related to data labeling. Therefore, there is a need to develop methods to minimize this while maintaining the quality of recognition at a acceptable level. One of the most popular methods for achieving this is active learning [18]. In classification, the most common is using previously labeled sets, with a decision determined based on observations of the object's features [9]. The methods used in active learning provide an opportunity to select the most representative examples for building models [6].

The study aims to develop data labeling methods based on active learning and train the model in a spiral annotation loop such that a model learned on partially labeled data is not statistically inferior to a model trained on a fully labeled set. Thanks to this, we expect that the amount of data required for manual labeling will be significantly reduced.

P. Trajdos and R. Burduk (Eds.): CORES 2025, LNNS 1557, pp. 93–102, 2025.
https://doi.org/10.1007/978-3-032-01773-4_10

In a nutshell, the main contributions of this work include

- developing new active learning strategies,
- comparison of the developed method with the benchmark strategies,
- performing a series of experiments using different deep neural network architectures.

2 Related Works

2.1 Active Learning

Active learning methods [12] can be divided into those based on uncertainty and probability distribution. Various metrics are defined for uncertainty-based methods to select the most uncertain samples based on the current model. In contrast, methods based on probability distributions aim to select samples based on their probability distributions.

Nowadays, a thriving, active learning field is called deep active learning of a given representation [3]. Active learning has applications in many different areas of classification and prediction [3], image segmentation [2], and many other areas. The aforementioned deep approach to active learning is based, among other things, on DAL networks, which are still being developed and feature different structures [10].

Many works showed that the classification quality after applying active learning methods could outperform traditional machine learning methods [14]. Also, in deep learning, applying AL methods gives promising results [1]. At the same time, we may face a problem related to the uncertainty associated with, for example, reducing the dataset needed to learn the network and preserve its discriminatory power. Different methods exist for assessing uncertainty for various combinations of network architecture. For example, some studies consider different budgets for acquiring data determination, low and high-budget strategies based on deep learning methods [8].

Many methods are used to label data under certain constraints, and one of them is to poll the oracle for the label depending on the distance from the decision boundary [16]. Another technique is Ward's hierarchical clustering, which aims to build a cluster structure to save as many labels as possible. For this purpose, unsupervised hierarchical clustering of unlabeled data or adaptive sampling is used [4].

2.2 Sampling Strategies

First, let us describe a few of the most popular sampling strategies [17]. Least Confidence (LC) involves returning samples for which the prediction score for the most likely class is the lowest. The LC value may be understood as the model's confidence that it will mislabel a given sample. Therefore, selecting samples with a high value of the LC measure for training can help give the model more confidence in the heaviest samples and thus significantly improve the model's performance. Margin Sampling (MS) consists of returning samples whose difference in prediction results between the two most likely classes is the smallest. The idea behind this method is that samples with a significant difference value are simple for the model, as it has no problem distinguishing between the most certain classes. On the other hand, samples with a slight difference in prediction results between the two most likely classes are complex, and selecting these for the training process can positively affect the model's performance. Entropy Sampling (ES) is based on returning samples whose entropy of the distribution of prediction results is the highest. It presumes that the higher the entropy, the more information a sample carries for the model being trained and the more advantageous it will be to select a given sample for the training. The most straightforward sampling is Random Sampling (RS) which returns samples in a pseudo-random fashion, regardless of the uncertainty of the prediction.

2.3 Active Learning Strategies

Various active learning strategies have been proposed [13]. Cumulative sample tuning involves running an iterative loop each time a newly labeled dataset appears, which involves tuning the model on all previously labeled data. Since the new samples in each iteration are selected using an acquisition function, in subsequent iterations, the dataset on which the model is tuned is expanded to include problematic and challenging ones for the model to label correctly. This approach to the problem ensures adequate expansion of the model's knowledge without catastrophic forgetting since samples from previous iterations are also included in the training sets of subsequent iterations of the Spiral Annotation Loop process. Another strategy is to train from scratch on cumulative samples. In the case of this strategy, every time a newly labeled dataset appears, an iteration of the loop is launched, which involves training the model from scratch on all previously labeled data. The model weights from which the training process begins are defined at the beginning of the entire Spiral Annotation Loop process and loaded at the beginning of each iteration. The idea behind this strategy is to counteract overfitting in the trained model. In the case of fine-tuning the model on all previously labeled data, there is a risk that the model will overfit on samples that appear earlier in the Spiral Annotation Loop process, and the model parameters will get stuck in a specific local minimum from which they will not be able to escape. The train-from-scratch on cumulative samples method prevents this phenomenon, as the model is not fine-tuned in each iteration but is trained from certain base weights that still need to be adapted to the dataset processed within the Spiral Annotation Loop. Fine-tune on currently iterated samples; every time a newly labeled dataset appears, an iteration of the loop is launched, which consists of fine-tuning the model on the newly labeled dataset. The clear advantage of this method is its efficiency and speed, as the training sets in each iteration are small, which allows for quick retraining of the model on the newly labeled dataset. Furthermore, since the following training samples are chosen using an acquisition function, it is assumed that the model is retrained on the currently most challenging samples, ensuring stable knowledge expansion of the trained model in subsequent iterations of the Spiral Annotation Loop process. Another approach is learning without forgetting strategy, which allows fine-tuning the model on newly labeled data without losing prediction accuracy on previously used data sets. As the name suggests, the implementation of this strategy is based on the Learning without Forgetting Algorithm and it uses regularization of the model's loss function by adding the distillation loss. The distillation loss is calculated between the predictions of the model trained in the current iteration of the Spiral Annotation Loop process and the predictions of the model trained in the previous iteration. This approach allows the model to expand its knowledge of the information in the newly labeled dataset without significantly changing the model's weights, which were crucial for correctly annotating samples in previous data sets. This strategy has the advantage of fine-tuning the currently iterated sample strategy, such as the high training speed of the model. Still, it also prevents the negative phenomenon of catastrophic forgetting [11]. The algorithm commences with limited labeled data in the learning-from-the-start method. The algorithm then selects informative examples from the unlabeled data in an iterative process. This selection process uses a strategy to maximize the model's uncertainty reduction. After choosing the most instructive examples, the labeled data is used to update the model, and the process is repeated until the model performance reaches a satisfactory level. *Uncertainty sampling* is a technique that involves choosing instances with the most significant level of uncertainty or highest entropy for annotation. *Query-by-Committee* (QBC) is another approach in which a group of models is trained on a dataset with annotations. Committee members then compete to predict the labels of the unannotated data, and the most valuable cases are identified by measuring the discrepancies between committee members. In contrast, the information density approach involves selecting labeled instances that represent the unlabeled dataset's diversity. The algorithm identifies examples that point to the entire dataset to ensure that the model understands the data; it often uses a clustering operation to aggregate the data [20]. To summarize, the strategy for choosing an approach depends on the problem and the data characteristics.

Active incremental learning trains a model on a small initial set of labeled data and integrates new unlabeled data by choosing the most informative annotated instances. The *Pool-based Dynamic Sampling* method entails an algorithm selecting a set of unlabeled instances, which is then ranked according to their informativeness. The algorithm then chooses the most informative examples from the pool for labeling while continually updating the model with new labeled data and selecting a fresh collection of instances to annotate. *Dynamic Ensemble Selection* is a very similar method to QBC, but it works differently for new samples. As new labeled data comes in, the algorithm updates the model and selects a new pool of examples to label. *Bayesian active learning* employs a Bayesian model to approximate the uncertainty of the algorithm's predictions. The algorithm identifies the most informative instances to annotate based on the uncertainty estimates. The method continually updates the Bayes model and selects new examples for labeling as new labeled data is acquired. Using specialized models such as Bayes Spline Neural Networks, significant improvements to other AL approaches for images can be obtained [7].

3 Methods

An active learning loop has been proposed as part of the work on selecting a method for supporting the data labeling process, utilizing a selected acquisition method and learning strategy. Its main goal is to enable the user to annotate an n-element data set in less time than full manual labeling and to obtain a trained neural network model at the end of the dataset labeling.

In the first step, a pre-defined, randomly sampled subset of data is fully annotated by hand. Based on these samples, the model is pre-trained on this subset. Then, predictions are made on all remaining instances in the dataset to identify the most informative ones based on the chosen metric. The selected subset forms the next batch of data to be manually annotated, which is then added to the training set of the network.

A spiral annotation loop results from selecting a subset, labeling it, and fine-tuning the network model. The annotation process is performed for 50% of the dataset or until the model achieves satisfactory accuracy, after which the model makes predictions on the remaining part of the dataset, thus propagating the knowledge gained on half of the dataset and labeling the unlabeled instances. The propagation of annotations completes the task of data annotation support.

The components of the active learning method are (a) an artificial neural network model, (b) an acquisition method algorithm determining the uncertainty of predictions of individual unlabeled data instances, and (c) an active learning strategy algorithm responsible for fine-tuning the model.

4 Experimental Evaluation

We conducted a series of computer experiments to verify the quality of the proposed method. In the course of them, we intended to answer the following research questions

RQ1: What is the minimum budget, ensuring that the quality of the resulting model is not inferior to the model learned on a fully labeled set?
RQ2: How quickly do the various AL strategies converge satisfactorily?

4.1 Setup

Experimental Protocol. We compared the active learning methods based on stratified datasets using *Repeated Stratified K-Fold (5 repeats $\times$ 2 splits) cross-validation* [5]. F-test at the significance level 5% was used as a statistical test to assess whether the quality of the proposed model trained on datasets with 50% of labeled data is no worse than that of the model trained on fully labeled datasets [19]. We chose *Accuracy* as the widely used metric to evaluate all methods.

Datasets. We conduct experiments on three visual (BloodCells, ASL, Melanoma) and one audio (TESS) datasets summarized in Table 1. Bloodcells dataset contains 12500 images of four cell types: Eosinophil, Lymphocyte, Monocyte, and Neutrophil. ASL dataset consists of 87000 images of hand gestures presenting 29 different letters and symbols in American Sign Language. Melanoma dataset is a set of 10600 skins images with two classes: benign and malignant. TESS dataset is a collection 2800 of audio files with recordings of two actresses saying phrases with different emotions: anger, disgust, fear, happiness, pleasant surprise, sadness, and neutral.

Table 1. Datasets used in experiments.

Name	Modality	Count	Class num	Task description
BloodCells	Image	12500	4	Cell classification on Eosinophil, Lymphocyte, Monocyte, or Neutrophil
ASL	Image	87000	29	Classification on signs from American Sign Language
Melanoma	Image	10600	2	Classification of melanoma on benign and malignant
TESS	Audio	2800	7	Classification of emotion from audio sample

Used Methods. Experiments were conducted for various combinations of sampling strategy and active learning methods. Acquisition methods were used in the research, i.e.:

- Entropy (E).
- Least Confidence (LC),
- Margin Sampling (MS),
- Random Sampling (R),

For each database, a series of preliminary experiments were conducted to select the best classifier and active learning strategy from

- Fine-tune on cumulative samples,
- Train from scratch on cumulative samples,
- Fine-tune on currently iterated samples,
- Learning without forgetting.

As a result of these experiments, the fine-tuning of the cumulative samples strategy was selected for Melanoma and ASL datasets, and the Train from scratch on cumulative samples strategy was chosen for the BloodCells and TESS datasets. Additionally, the following models were used:

- ResNeXt50-32x4d for ASL,
- EfficientnetB0 for BloodCells,
- RESNET-18 for Melanoma,
- ResNext101 for TESS.

Implementation and Reproducibility. The methods and experimental environment are implemented using the Python programming language and the following libraries: scikit-learn [1], torchvision [2], NumPy [3], and pandas [4]. The source code (sufficient to repeat the experiments) and detailed experimental results are available upon request.

4.2 Results

The summary results of the experiment are included in Table 2, while the learning curves (dependence between accuracy and number of labeled objects) for each dataset are shown in Fig. 1. We updated the models during the experiments using 5% labeled examples in each step.

Table 2. Summary results of the experiments. 5% of the labeled objects were added to the learning set according to a sampling strategy in each cycle.

	DATA SETS															
	10BC				ASL				MEL				TESS			
	METHODS															
Cycle	E	LC	MS	R	E	LC	MS	R	E	LC	MS	R	E	LC	MS	R
0	0.242	0.242	0.242	0.242	0.010	0.010	0.010	0.010	0.365	0.365	0.365	0.365	0.038	0.038	0.038	0.038
1	0.651	0.643	0.655	0.652	0.108	0.141	0.417	0.484	0.653	0.637	0.639	0.823	0.187	0.258	0.273	0.534
2	0.868	0.856	0.864	0.837	0.332	0.452	0.633	0.639	0.857	0.757	0.818	0.824	0.543	0.763	0.597	0.864
3	0.921	0.929	0.914	0.866	0.508	0.648	0.728	0.716	0.871	0.863	0.873	0.867	0.800	0.967	0.733	0.972
4	0.945	0.939	0.949	0.894	0.648	0.723	0.749	0.753	0.868	0.876	0.887	0.863	0.989	0.889	0.922	0.995
5	0.954	0.945	0.955	0.927	0.741	0.761	0.725	0.750	0.873	0.903	0.906	0.868	0.998	0.993	0.993	0.742
6	0.962	0.957	0.954	0.934	0.763	0.753	0.756	0.758	0.909	0.899	0.901	0.852	0.930	1.000	1.000	0.982
7	0.964	0.963	0.934	0.937	0.766	0.765	0.761	0.766	0.913	0.911	0.912	0.902	1.000	0.987	1.000	0.988
8	0.962	0.960	0.966	0.932	0.768	0.766	0.769	0.757	0.913	0.911	0.934	0.889	0.998	0.998	0.996	0.992
9	0.959	0.965	0.961	0.936	0.764	0.764	0.774	0.767	0.924	0.925	0.932	0.898	0.998	1.000	0.996	0.998
10	0.942	0.958	0.961	0.955	0.769	0.742	0.770	0.748	0.927	0.928	0.920	0.877	1.000	1.000	0.963	0.996
11	0.966	0.960	0.965	0.946	0.767	0.768	0.645	0.731	0.927	0.929	0.931	0.908	1.000	0.991	1.000	1.000
12	0.952	0.959	0.965	0.953	0.768	0.768	0.767	0.754	0.924	0.933	0.932	0.913	1.000	0.993	1.000	1.000
13	0.966	0.963	0.966	0.955	0.768	0.765	0.756	0.758	0.927	0.941	0.934	0.924	1.000	1.000	1.000	0.996
14	0.965	0.964	0.965	0.962	0.767	0.765	0.770	0.744	0.938	0.923	0.930	0.923	1.000	1.000	1.000	0.928
15	0.963	0.965	0.965	0.961	0.765	0.765	0.771	0.763	0.935	0.947	0.940	0.920	0.998	0.998	1.000	1.000
16	0.962	0.965	0.968	0.952	0.766	0.768	0.763	0.765	0.924	0.939	0.946	0.939	1.000	0.983	1.000	1.000
17	0.968	0.965	0.966	0.967	0.752	0.765	0.769	0.767	0.947	0.940	0.941	0.939	1.000	0.997	1.000	0.996
18	0.968	0.967	0.966	0.960	0.767	0.764	0.722	0.768	0.936	0.937	0.940	0.917	1.000	0.998	0.981	1.000
19	0.970	0.968	0.945	0.967	0.768	0.767	0.767	0.764	0.937	0.938	0.944	0.929	1.000	1.000	1.000	1.000
20	0.959	0.969	0.968	0.968	0.767	0.767	0.697	0.760	0.932	0.942	0.945	0.945	-	-	-	-
MIN	0.242	0.242	0.242	0.242	0.010	0.010	0.010	0.010	0.365	0.365	0.365	0.365	0.038	0.038	0.038	0.038
MAX	0.970	0.969	0.968	0.968	0.769	0.768	0.774	0.768	0.947	0.947	0.946	0.945	1.000	1.000	1.000	1.000
MEAN	0.905	0.905	0.904	0.891	0.659	0.676	0.691	0.701	0.876	0.873	0.879	0.871	0.874	0.893	0.875	0.901
SD	0.167	0.168	0.167	0.164	0.227	0.213	0.176	0.171	0.133	0.138	0.136	0.121	0.283	0.263	0.270	0.234

E – ENTROPY; LC - LEAST CONFIDENCE; MS - MARGIN SAMPLING; R – RANDOM; SD - standard deviation

We compared the results achieved for models trained on 50% labeled datasets with the quality of models trained on fully labeled datasets using a 5×2cv F-test

[1] https://scikit-learn.org/stable/.
[2] https://pytorch.org/vision/stable/index.html.
[3] https://numpy.org/.
[4] https://pandas.pydata.org/.

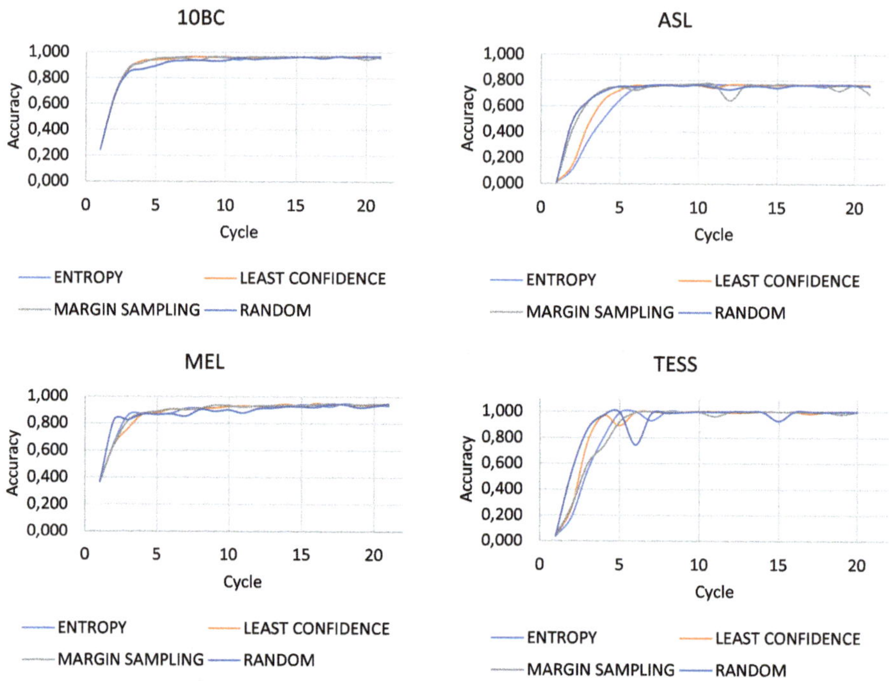

Fig. 1. Relationship between accuracy and number of labeled objects for tested datasets - Summary results of the experiments. 5% of the labeled objects were added to the learning set according to a sampling strategy in each cycle.

for a confidence level of 5%. For all cases, the accuracy of models learned using active learning was not statistically significantly worse than the reference models, and the p-values for individual datasets were <0.001 for ASL and BloodCells, 0.011 for Melanoma, and 0.038 for TESS.

4.3 Lessons Learned

Let's try to answer the research questions asked at the beginning.

RQ1: What is the minimum budget, ensuring that the quality of the resulting model is not inferior to the model learned on a fully labeled set?

We showed that 50% of labeled objects is sufficient to get a model with no statistically worse accuracy than a model based on a fully labeled dataset. However, in most cases, the requirement was much smaller. As we can see, the more difficult the problem was, the higher the requirement for the number of labeled objects was. For BloodCells or Melanoma, less than 25% was sufficient for each strategy, while 30 to 40% was required for ASL and TESS collections. RQ1 answered.

RQ2: How quickly do the various AL strategies converge satisfactorily?

As described above, the proposed methods reach convergence fairly quickly, and adding more labeled objects no longer improves their quality. It is also worth noting that the best sampling strategy cannot be identified. For the Melanoma and TESS, it was Entropy, however, for BloodCels, the best method was Margin Sampling. For ASL, both Margin Sampling and Random Sampling yielded the best results. This means that for practical use, the sampling method should be treated as an important hyperparameter that must be carefully selected for a given problem. RQ2 answered.

5 Conclusion

This work aimed to analyze the chosen characteristics of active learning strategies and the sampling methods for deep classification models. As shown, for selected decision tasks in the field of image and audio signal classification, it was possible to choose methods that, on the one hand, did not require a large volume of labeled data to achieve quality, not worse than models trained on a fully labeled dataset. Additionally, the convergence of models is relatively fast. The presented experimental results are promising and encourage using the obtained solutions for classification problems in the continual learning scenario, especially when there may be changes in the incoming data distribution. The analyzed sampling methods can also be helpful for another scenario when new classes should be trained (class incremental learning), and to prevent catastrophic forgetting, data buffers are used to train a model on objects from previously learned classes periodically. The buffer lineup could be determined using, e.g., the mentioned sampling methods. However, the paper's main achievement is analyzing the usefulness of the model's retraining strategies after adding new objects to the learning set according to a chosen sampling method. It was shown that for some tasks the often-used *train from scratch on cumulative samples* strategy can be successfully replaced by strategies allowing incremental learning of the model. Strategies based on fine-tuning were used for this. It is also worth mentioning that the approach based on continual learning, i.e., *learning without forgetting*, returns slightly worse results than the above mentioned strategies.

Acknowledgement. This work was supported by the National Centre for Research and Development under the Smart Growth Operational program, contract number POIR.01.01.01-00-0244/21.

References

1. Beluch W H, Genewein T, Nürnberger A, K.J.M.: The power of ensembles for active learning in image classification. In Proceedings of the IEEE Conference on Computer Vision and Pattern Recognition, pp. 9368–9377 (2018)
2. Casanova A, Pinheiro P O, Rostamzadeh N, P.C.: Reinforced active learning for image segmentation. arXiv preprint arXiv:2002.06583 (2020)

3. Chen, Y., Lasko, T.A., Mei, Q., Denny, J.C., Xu, H.: A study of active learning methods for named entity recognition in clinical text. J. Biomed. Inf. **58**, 11–18 (2015)
4. Dasgupta, D.H.S.: Hierarchical sampling for active learning. ICML Helsinki (2008)
5. Demšar, J.: Statistical comparisons of classifiers over multiple data sets. J. Mach. Learn. Res. **7**, 1–30 (2006)
6. Fu, B.L.Y.: A survey on instance selection for active learning. Technical Report, University of Wisconsin, Madison (TR1648), pp. 1–35 (2013)
7. Gal, Y., Islam, R., Ghahramani, Z.: Deep bayesian active learning with image data (2017)
8. Hacohen G, Dekel A, W.D.: Active learning on a budget: opposite strategies suit high and low budgets. Cornell University. arXiv:2202.02794v4 [cs.LG] (2022)
9. Hastie, T., Tibshirani R, Friedman, J.H.: The elements of statistical learning: data mining, inference and prediction. Math. Intelligencer **27(2)**, 83–85 (2005)
10. Jordan, T., Zhang, C., Krishnamurthy, A., Langford, J., Agarwal, A.: Deep batch active learning by diverse, uncertain gradient lower bounds. In: 8th International Conference on Learning Representations, ICLR 2020, Addis Ababa, Ethiopia, April 26–30, 2020. OpenReview.net (2022)
11. Li, Z., Hoiem, D.: Learning without forgetting. arXiv:1606.09282 (2016)
12. Gao, M., Zhang, Z., Yu, G., Arık, S.Ö., Davis, L.S., T, P.: Statere labeling adversarial active learning. In: Proceedings of the IEEE/CVF Conference on Computer Vision and Pattern Recognition, Seattle, WA, USA, pp. 8756–8765 (2020)
13. Margatina K, Barrault L, A.N.: On the importance of effectively adapting pre-trained language models for active learning. arXiv:2104.08320v2 [cs.CL] (2022)
14. Naseem, J.U., Khushi, M., Khalid, S., Shaukat, K., M, M.A.: A comparative analysis of active learning for biomedical text mining. Appl. Syst. Innov. **4(1)** (2021)
15. Nguyen, T., Gosine, R. , W, P.: Asystematic review of big data analytics foroil and gas industry4.0. IEEE Access **8**, 183–201 (2020). https://doi.org/10.1109/ACCESS.2020.2979678
16. Schutze, H., Velipasaoglu, E., Pedersen, J.: Performance thresholding in practical text classification. In: ACM International Conference on Information and Knowledge Management (2006)
17. Settles, B.: Active learning literature survey. Computer Sciences Technical Report 1648, University of Wisconsin–Madison (2010)
18. Settles, B.: Active learning literature survey. https://minds.wisconsin.edu/handle/1793/60660. Accessed 31 Mar 2023, pp. 73–78 (2019)
19. Stapor, K., Ksieniewicz, P., García, S., Woźniak, M.: How to design the fair experimental classifier evaluation. Appl. Soft Comput. **104**, 107219 (2021). https://doi.org/10.1016/j.asoc.2021.107219, https://www.sciencedirect.com/science/article/pii/S1568494621001423
20. Wang, M., Min, F., Zhang, Z.H., Wu, Y.X.: Active learning through density clustering. Expert Syst. Appl. **85**, 305–317 (2017)

A Prototype of an AI-Driven Operational Security System for FinTechs: A FaaS-Based Approach to Fraud Detection

Andrzej Dorochowicz[3](✉), Dariusz Jankowski[1], Paweł Ksieniewicz[1], Katarzyna Topolska[2], Mariusz Topolski[1], and Paweł Zyblewski[1]

[1] Department of Systems and Computer Networks, Faculty of Information and Communication Technology, Wrocław University of Science and Technology, Wrocław, Poland
{dariusz.jankowski,pawel.ksieniewicz,mariusz.topolski, pawel.zyblewski}@pwr.edu.pl

[2] Department of Control Systems and Mechatronics, Faculty of Information and Communication Technology, Wrocław University of Science and Technology, Wrocław, Poland
katarzyna.topolska@pwr.edu.pl

[3] INCAT Sp. z o.o., Braniborska 40, 53-680 Wrocław, Poland
andrzej.dorochowicz@incat.eu
https://incat.eu/ , https://www.kssk.pwr.edu.pl/

Abstract. This paper presents an operational security platform for financial institutions, built on advanced machine learning algorithms and offered in a Fintech-as-a-Service (FaaS) model. The system integrates transactional layers with external AML, KYC, and fraud detection services, enabling real-time anomaly detection. A comparative evaluation of selected classification algorithms was conducted, employing various resampling, dimensionality reduction, and classifier ensemble strategies. Gaussian Naive Bayes demonstrated superior accuracy and efficiency in probabilistic risk estimation, making it optimal for real-time deployment. The platform's modularity, integration layer, and hybrid classification engine offer high adaptability to evolving financial threat landscapes.

Keywords: FinTech Security Automation · Fraud Detection Algorithms · Ensemble Learning · Difficult data

1 Introduction

In the rapidly evolving financial sector, ensuring operational security has become paramount [1]. The development of an innovative operational security platform based on advanced machine learning mechanisms offered in the Fintech as a Service (FaaS) model addresses the increasing challenges faced by financial institutions [2]. This platform, primarily designed for FinTech organizations and Challenger Banks, integrates transactional systems with external operational service

P. Trajdos and R. Burduk (Eds.): CORES 2025, LNNS 1557, pp. 103–112, 2025.
https://doi.org/10.1007/978-3-032-01773-4_11

providers, including Anti-Money Laundering (AML) [3], Know Your Customer (KYC) [4], card services [5], scoring [6], fraud detection [7], and clearing systems [8]. Integrating artificial intelligence (AI) into the intermediary layer facilitates the automation of operational processes and minimizes operational risks [9]. Operational security encompasses four fundamental areas: operational continuity [10], data security [11], transaction security [12], and contract security [13]. The focus of the research is on transaction security and contract security; however, automating operational processes also impacts operational continuity, particularly concerning the business efficiency of the solution [14]. Implementing AI mechanisms in operational processes enables the development of algorithms capable of eliminating or significantly reducing human involvement in verifying the security of financial operations [15]. Previous studies have verified the feasibility of using point interfaces to external systems involved in operational processes [16]. This verification included the possibility of parameterizing interfaces, constructing variant queries, and technological aspects related to the direct use of existing APIs [17]. The conducted research and literature studies have confirmed the possibility of variant query construction and decision path development [18], the application of machine learning algorithms to detect fraud and abuses in banking transactions, including fraud detection, scoring, and money laundering [19], as well as the use of adaptive methods that allow predictive models to adjust to changing class distributions over time, associated, for example, with the development of new fraud mechanisms [20]. This indicates the existence of confirmed mechanisms enabling the development of effective classification algorithms for individual tasks related to the operational security of banking systems.

The aim of the research was to develop and validate an AI-driven operational security platform for financial entities, particularly FinTechs and Challenger Banks, using advanced machine learning mechanisms in a Fintech-as-a-Service (FaaS) model. The system integrates transactional layers with external services (AML, KYC, scoring, fraud detection, etc.) via an AI-powered intermediary layer to automate operational processes and reduce risk [21,22]. Focusing on transaction and contract security, the platform also contributes to operational continuity by minimizing human intervention. The research involved verifying the feasibility of integrating external systems via parameterizable APIs and constructing adaptive, variant decision paths using ML classifiers for fraud detection and risk assessment [21–24,26–28]. Emphasis was placed on adaptive models capable of adjusting to data drift over time [25]. To reach high technological readiness, industrial research developed predictive components and a hybrid classification system using real financial data [27,28], followed by experimental deployment and validation of the Incat FaaS AI platform in a real-world production environment. The core research goal was to compare classifier performance on experimental and live operational data.

2 Methodology

2.1 Datasets

The dataset comprises 35,165 records related to transaction evaluations across eight financial processes designed to enhance operational security:

1. **Know Your Customer (KYC)**: Verifies customer identity and credibility, aiming to block suspicious transactions, prevent identity theft, and support AML compliance. Operator validation is based on selected attributes during onboarding.
2. **Anti-Money Laundering (AML)**: Ensures compliance with EU Directive 2015/849 by continuously analyzing transactions for suspicious activity. Final decisions are operator-driven based on predefined verification criteria.

3 & 4. **Transaction AML (Internal/External)**: Evaluate transaction legitimacy using internal or external customer data. Operators assess attributes and intervene in ambiguous or high-risk cases.

5 & 6. **Card Transaction AML (Internal/External)**: Assess card transactions for anomalies. Decisions depend on operator evaluation of defined features, with escalation when additional scrutiny is needed.

7. **Credit Card Application**: Determines applicant creditworthiness using data such as credit score, income, employment type, and debt ratios. Legal and financial criteria must be met.
8. **Overdraft Application**: Evaluates customer eligibility for overdraft limits based on account activity, income evidence, credit history, and employment data.

In this process-level analysis, the system assumes that new features may dynamically emerge, requiring continuous risk assessment based on both internal and external customer data. A binary classification approach was applied. The results presented in this work refer to a prototype system, tested mostly in a static environment, while the complete system, prepared for deployment, also took into account the data stream processing.

2.2 Evaluation of Classification Accuracy and Adaptability

Figure 1 presents a comparison between the performance of predictive models built on historical data and their effectiveness when applied to real-world data no older than six months. The evaluation is based on the Balanced Accuracy Score (BAC), which accounts for both sensitivity and specificity, making it particularly suitable for assessing models trained on imbalanced datasets. Eight business processes are analyzed, including procedures related to anti-money laundering (AML), customer verification, and banking product operations. Each bar represents the mean BAC with a 95% confidence interval.

For the Know Your Customer (KYC) process, the model shows nearly perfect alignment between predictions on historical data and current data, indicating high model stability and consistent data characteristics over time. A similar

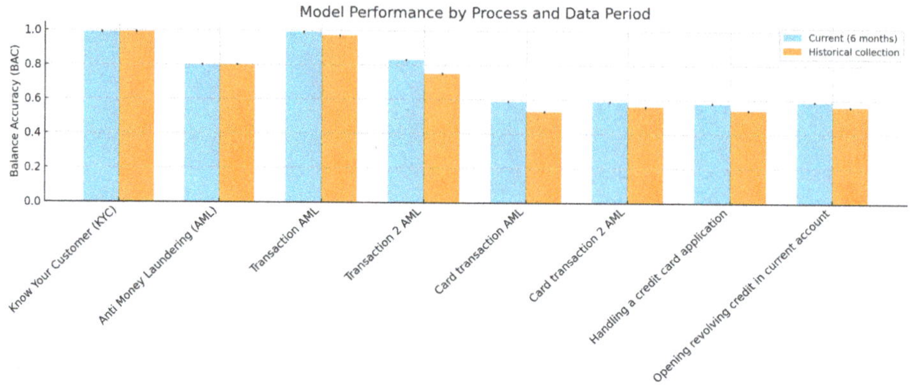

Fig. 1. Results obtained in the first experiment.

trend can be observed in the Transaction AML process, where BAC remains high for both datasets, although a slight decrease in performance with the recent data is noticeable. In contrast, processes such as Transaction 2 AML and Card Transaction AML show a more pronounced drop in BAC for current data, suggesting changes in input data patterns or a lack of timely model updates. Even greater performance declines are observed in Card Transaction 2 AML, Handling a Credit Card Application, and Opening Revolving Credit in Current Account, pointing to a potential degradation in data quality, changes in customer behavior, or the need for model retraining with recent data. Among all processes, models related to card transactions and credit products appear to be the most sensitive to data shifts. This may be due to evolving market conditions, seasonal behavior, or emerging user activity patterns that differ from the historical trends on which the models were trained. The analysis reveals that while some models maintain high predictive accuracy over time, others show significant performance drops when applied to recent data. This highlights the need for regular recalibration or retraining of models to reflect new data realities. Particular attention should be given to AML processes involving card transactions and credit products, where the largest BAC declines were observed. Continuous model monitoring and timely updates are essential to ensure sustained performance and operational reliability.

2.3 Classifier Committees and Resampling Techniques

In the following experiments, a broad set of classification algorithms, resampling methods, feature extractors, and decision aggregation rules was evaluated. Tested classifiers included Multilayer Perceptron (MLP), Gaussian Naive Bayes (GNB), Support Vector Classifier (SVC), CART, k-Nearest Neighbours (kNN), TabNet, and CEM (a feedforward neural network with embeddings). To address the class imbalance, SMOTE, Random Oversampling (ROS), Random Undersampling (RUS), and no resampling (NONE) were applied. Feature reduction techniques included PCA, IPCA, FA, ICA, NMF, tSVD, LDA, and a no-transformation baseline (Noext).

Ensemble strategies such as Bagging and Random Subspace were used to enhance generalization. At the same time, decision fusion rules—Min, Max, Product (Pro), and Majority Voting (MVo)—were tested to optimize classifier committee performance. The pipeline combined preprocessing, feature extraction, and classification steps to assess their joint impact on accuracy. Gaussian Naive Bayes achieved the best performance and was selected for final evaluations.

The experimental software consists of two modules: an experimental module (Fig. 2) that performs data loading, 5 times repeated 2-fold Cross-Validation, and model training using all preprocessing:extraction:classification combinations, and an analytical module that computes metrics, performs statistical analyses, and outputs comprehensive reports. All models and results are stored in serialized formats (pickle, NumPy). The pipeline design enables systematic benchmarking of ML workflows for financial anomaly detection.

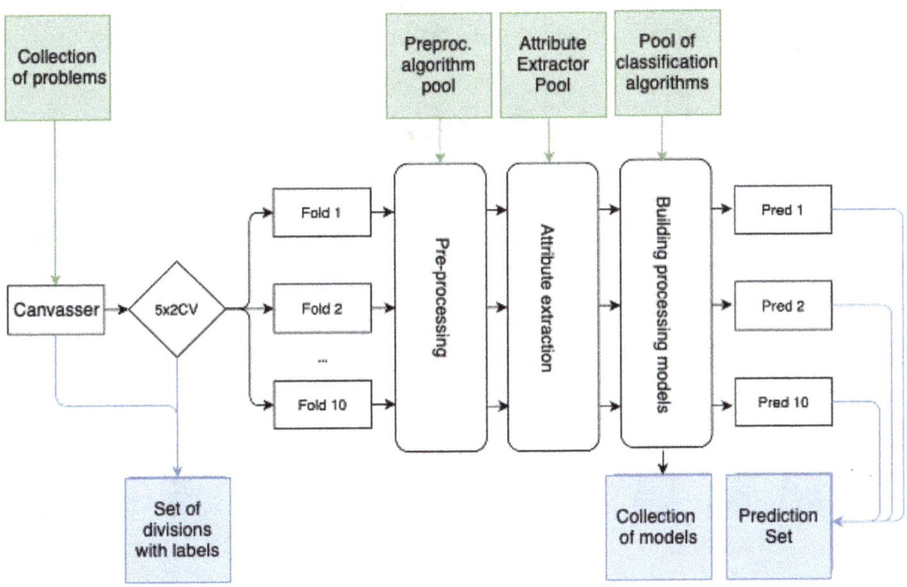

Fig. 2. Schema the experimental evaluation software project.

As part of the study, classifier committees were evaluated to improve overall classification performance beyond that of individual models. A key research question involved determining the optimal number of classifiers and how their outputs should be aggregated. The following decision functions were tested: **Min**, **Max**, **Product (Pro)**, and **Majority Voting (MVo)**.

To enhance classifier diversity, sampling-based ensemble methods were applied. **Bagging** generated multiple training subsets by resampling the data, training identical algorithms on each. **Random Subspace** instead sampled feature subsets for each model, which is effective in high-dimensional settings.

Various resampling strategies were also used to handle class imbalance:

- NONE - preserved the original dataset.
- SMOTE - generated synthetic minority samples using interpolation.
- ROS - duplicated minority class samples randomly, increasing risk of overfitting.
- RUS - removed majority class samples, which may reduce training data quality.

These methods supported the development of robust committee models for transaction-level classification tasks.

2.4 Offline Evaluation in Experimental Environment

Based on the analysis of the presented Fig. 3, the best classification results were achieved using different configurations depending on the number of classes and the combination of diversification, integration, and resampling methods.

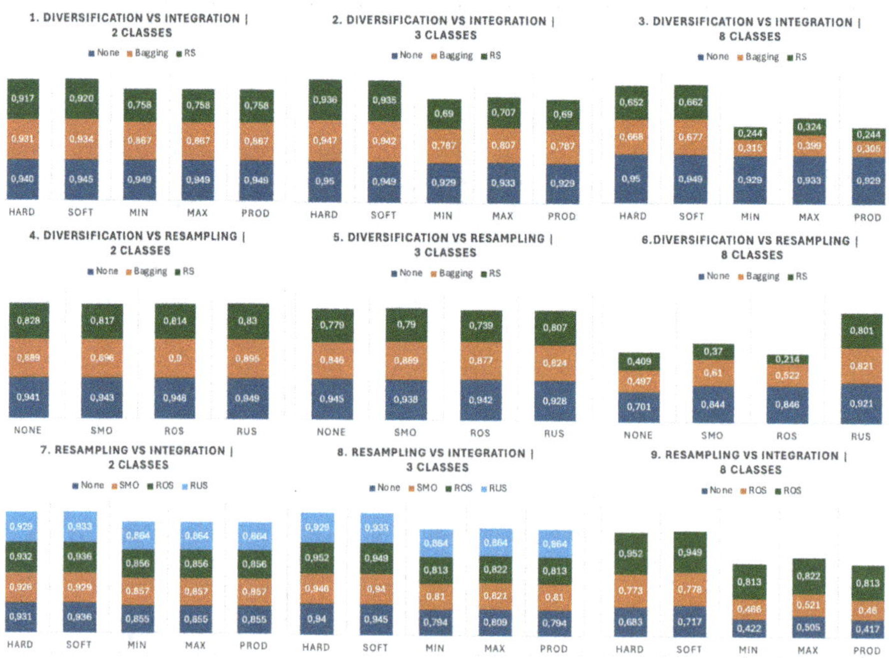

Fig. 3. Results obtained in the first experiment.

For binary classification problems (2 classes), the highest result (0.949) was achieved using integration methods MIN, MAX, and PROD without additional diversification. An equally high result was observed when combining diversification with resampling (RUS), again without applying integration. In the comparison between resampling and integration, the best result (0.933) was achieved with Soft integration and either no resampling or SMO.

In three-class problems, the best result (0.952) was obtained using the ROS resampling method combined with Hard integration. Another strong result (0.950) was observed with Hard diversification and no integration. In the diversification vs. resampling comparison, the highest performance (0.945) was again achieved without applying any resampling or integration methods.

In eight-class problems, the top results (0.952) were obtained both with no resampling and Hard integration, and with ROS resampling combined with Hard integration. It is also worth noting that a result of 0.950 was achieved with Hard diversification and no integration.

In summary, optimal classification results were achieved with HARD integration or NONE, MIN or ROS/RUS resampling, and HARD diversification—particularly in multi-class scenarios.

2.5 Deployment in Real FaaS Environment

Deploying machine learning models involves transitioning from development to production environments, each using distinct tools and workflows. In development, models are built and tested using programming libraries and utilities. In production, models must run reliably in real-time, solving practical business tasks.

The system architecture, shown in Fig. 4, supports this transition.

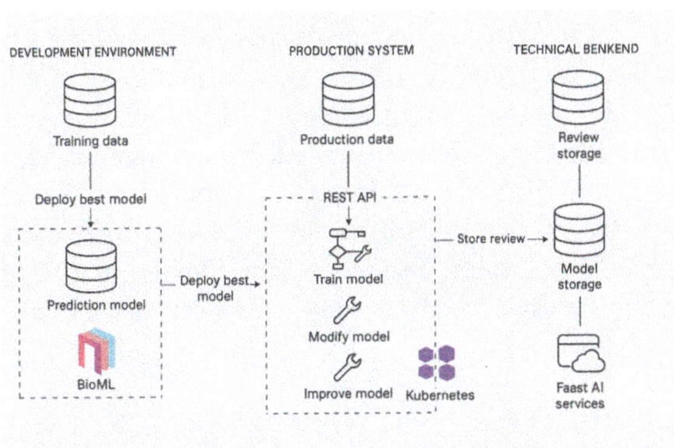

Fig. 4. AI Architecture for BOS_FaaS.

Key components include:

- Training and Production Data - Sourced from eight FaaS AI processes, used for both model training and real-time input.

- Prediction Model - Forecasts outcomes based on input variables.
- Kubernetes & BentoML - Handle containerized model deployment and management.
- Data & Model Repositories - Store unstructured data and model containers.
- FaaS AI Services - Consume production data and trigger processing logic.

This phase validates the FaaS AI system in production-like conditions. Tests confirmed high classification performance, with the Naive Bayes model yielding the best results, demonstrating readiness for integration into the BOS platform.

3 Conclusion

This study presents a comprehensive approach to enhancing operational security in financial institutions through an AI-driven platform delivered in the FinTech-as-a-Service (FaaS) model. The system integrates transactional infrastructure with external services such as AML, KYC, fraud detection, and scoring, enabling real-time decision-making through advanced machine learning mechanisms.

A key contribution of the research is the development and evaluation of a hybrid classification architecture that leverages diverse ML models, ensemble methods, and data resampling strategies. Extensive offline experiments demonstrated that the Gaussian Naive Bayes classifier consistently outperformed others in terms of both accuracy and computational efficiency, particularly when combined with Hard integration and selective resampling (ROS/RUS).

The platform was validated in both experimental and real-world environments, including deployment in the Microsoft Azure cloud using containerized services (Kubernetes, BentoML). Production-level tests confirmed that the system operates reliably under real-time conditions and achieves classification quality comparable to or better than that obtained in controlled offline scenarios.

The results indicate that the proposed solution is technologically mature, modular, and adaptable to the evolving needs of FinTech organizations and Challenger Banks. It provides a scalable foundation for automated, intelligent transaction verification, contributing to increased operational continuity, reduced fraud risk, and improved data-driven decision-making in financial services.

The research concerns Incat FaaS AI âĂŞ Development of an operational security platform for financial institutions based on advanced machine learning mechanisms.

Acknowledgement. This work was supported by the National Centre for Research and Development under the Smart Growth Operational program, contract number POIR.01.01.01-00-0885/19-00.

References

1. Pavlidis, G.: Deploying artificial intelligence for anti-money laundering and asset recovery: the dawn of a new era. J. Money Laundering Control **26**(7), 155–166 (2023). https://doi.org/10.1108/JMLC-03-2023-0050

2. Ngai, E., Hu, Y., Wong, Y.H., Chen, Y., Sun, X.: The application of data mining techniques in financial fraud detection: a classification framework and an academic review of literature. Decis. Support Syst. **50**(3), 559–569 (2011). https://doi.org/10.1016/j.dss.2010.08.006
3. Fawcett, T., Provost, F.: Adaptive fraud detection. Data Min. Knowl. Disc. **1**(3), 291–316 (1997). https://doi.org/10.1023/A:1009700419189
4. Van Vlasselaer, V., Eliassi-Rad, T., Akoglu, L., Snoeck, M., Baesens, B.: APATE: a novel approach for automated credit card transaction fraud detection using network-based extensions. Decis. Support Syst. **75**, 38–48 (2015). https://doi.org/10.1016/j.dss.2015.04.013
5. Carcillo, F., Le Borgne, Y.A., Caelen, O., Bontempi, G.: Streaming active learning strategies for real-life credit card fraud detection: assessment and visualization. Int. J. Data Sci. Anal. **5**(4), 285–300 (2018)
6. Fiore, U., De Santis, A., Perla, F., Zanetti, P., Palmieri, F.: Using generative adversarial networks for improving classification effectiveness in credit card fraud detection. Inf. Sci. **479**, 448–455 (2019). https://doi.org/10.1016/j.ins.2017.12.030
7. Bahnsen, A.C., Aouada, D., Ottersten, B.: Example-dependent cost-sensitive decision trees. Expert Syst. Appl. **42**, 6609–6619 (2016). https://doi.org/10.1016/j.eswa.2015.04.042
8. Colladon, A.F., Remondi, E.: Using social network analysis to prevent money laundering. Expert Syst. Appl. **67**, 49–58 (2017). https://doi.org/10.1016/j.eswa.2016.09.029
9. Adedotun, O., Olanrewaju, O.O.: Enhancing cybersecurity in fintech: safeguarding financial data against evolving threats and vulnerabilities. Int. J. Comput. Appl. Technol. Res. **14**(1), 62–68 (2025). https://doi.org/10.7753/IJCATR1401.1005
10. Eryu, P.: Machine learning in financial transaction fraud detection and prevention. Trans. Econ. Bus. Manag. Res. **5**, 243–249 (2024). https://doi.org/10.62051/16r3aa10
11. Jumayeva, Z.: Ensuring economic security in the banking sector. Gospodarka i Innowacje **47** (2024)
12. Conti, M., Kumar, M., Lal, C., Ruj, S.: A survey on security and privacy issues of Bitcoin. IEEE Commun. Surv. Tutorials **20**(4), 3416–3452 (2018). https://doi.org/10.1109/COMST.2018.2842460
13. Zetzsche, D.A., Buckley, R.P., Arner, D.W., Barberis, J.N.: From FinTech to TechFin: the regulatory challenges of data-driven finance. N. Y. Univ. J. Law Bus. **14**(2), 393–446 (2017)
14. Doumpos, M., Zopounidis, C.: A multicriteria decision support system for bank rating. Decis. Support Syst. **50**, 55–63 (2010). https://doi.org/10.1016/j.dss.2010.07.002
15. Goodell, J.W., Goutte, S.: COVID-19 and finance: agendas for future research. Financ. Res. Lett. **35**, 101512 (2020). https://doi.org/10.1016/j.frl.2020.101512
16. King, B.: Bank 4.0: Banking everywhere, never at a bank. Wiley (2018)
17. Böhme, R., Christin, N., Edelman, B., Moore, T.: Bitcoin: economics, technology, and governance. J. Econ. Perspect. **29**(2), 213–238 (2015). https://doi.org/10.1257/jep.29.2.213
18. Mukesh, A., Sofia, R.: AI-driven risk management in financial markets and fintech. IEEE (2024). https://doi.org/10.1109/IC3TES62412.2024.10877453
19. Jagtiani, J., Lemieux, C.: The roles of alternative data and machine learning in fintech lending: evidence from the US market. Financ. Manage. **48**(4), 1009–1029 (2019). https://doi.org/10.1111/fima.12295

20. Awosika, T., Shukla, R.M., Pranggono, B.: Transparency and privacy: the role of explainable AI and federated learning in financial fraud detection. IEEE Access **12**, 64551–64560 (2024). https://doi.org/10.1109/ACCESS.2024.3394528
21. Abellán, J., Castellano, J.G.: A comparative study on base classifiers in ensemble methods for credit scoring. Expert Syst. Appl. **73**, 1–10 (2018)
22. Bahnsen, A.C., Stojanovic, A., Aouada, D., Ottersten, B.: Improving credit card fraud detection with calibrated probabilities. In: SIAM International Conference on Data Mining 2014 (SDM 2014), pp. 677–685 (2014)
23. Chen, Z., Van Khoa, L.D., Teoh, E.N., Nazir, A., Karuppiah, E.K., Lam, K.S.: Machine learning techniques for anti-money laundering (AML) solutions in suspicious transaction detection: a review. Knowl. Inf. Syst. **57**(2), 245–285 (2018). https://doi.org/10.1007/s10115-017-1144-z
24. Dreżewski, R., Sepielak, J., Filipkowski, W.: System supporting money laundering detection. Digit. Investig. **9**(1), 8–21 (2012)
25. Krawczyk, B., Minku, L.L., Gama, J., Stefanowski, J., Woźniak, M.: Ensemble learning for data stream analysis: a survey. Inf. Fusion **37**, 132–156 (2017)
26. Luo, X.: Suspicious transaction detection for anti-money laundering. Int. J. Secur. Appl. **8**(2), 157–166 (2016)
27. Wang, D., Chen, B., Chen, J.: Credit card fraud detection strategies with consumer incentives. Omega **88**, 179–195 (2019)
28. Zelenkov, Y.: Example-dependent cost-sensitive adaptive boosting. Expert Syst. Appl. **135**, 71–82 (2019)

Author Index

P. Trajdos and R. Burduk (Eds.): CORES 2025, LNNS 1557, p. 113, 2025.
https://doi.org/10.1007/978-3-032-01773-4

GPSR Compliance
The European Union's (EU) General Product Safety Regulation (GPSR) is a set of rules that requires consumer products to be safe and our obligations to ensure this.

If you have any concerns about our products, you can contact us on

ProductSafety@springernature.com

In case Publisher is established outside the EU, the EU authorized representative is:

Springer Nature Customer Service Center GmbH
Europaplatz 3
69115 Heidelberg, Germany

www.ingramcontent.com/pod-product-compliance
Lightning Source LLC
Chambersburg PA
CBHW050917190825
31302CB00030B/627